Crowd Dynamics
by Kinetic Theory Modeling

Complexity, Modeling, Simulations, and Safety

Synthesis Lectures on Mathematics and Statistics

Editor
Steven G. Krantz, *Washington University, St. Louis*

Introduction to Statistics Using R
Mustapha Akinkunmi
2019

Inverse Obstacle Scattering with Non-Over-Determined Scattering Data
Alexander G. Ramm
2019

Analytical Techniques for Solving Nonlinear Partial Differential Equations
Daniel J. Arrigo
2019

Aspects of Differential Geometry IV
Esteban Calviño-Louzao, Eduardo García-Río, Peter Gilkey, JeongHyeong Park, and Ramón Vázquez-Lorenzo
2019

Symmetry Problems. The Navier–Stokes Problem.
Alexander G. Ramm
2019
An Introduction to Partial Differential Equations
Daniel J. Arrigo
2017

Numerical Integration of Space Fractional Partial Differential Equations: Vol 2 – Applicatons from Classical Integer PDEs
Younes Salehi and William E. Schiesser
2017

Numerical Integration of Space Fractional Partial Differential Equations: Vol 1 – Introduction to Algorithms and Computer Coding in R
Younes Salehi and William E. Schiesser
2017

Aspects of Differential Geometry III
Esteban Calviño-Louzao, Eduardo García-Río, Peter Gilkey, JeongHyeong Park, and Ramón Vázquez-Lorenzo
2017

The Fundamentals of Analysis for Talented Freshmen
Peter M. Luthy, Guido L. Weiss, and Steven S. Xiao
2016

Crowd Dynamics by Kinetic Theory Modeling: Complexity, Modeling, Simulations, and Safety

Bouchra Aylaj, Nicola Bellomo, Livio Gibelli, and Damián Knopoff

ISBN: 978-3-031-01300-3   paperback
ISBN: 978-3-031-02428-3   ebook
ISBN: 978-3-031-00274-8   hardcover

DOI 10.1007/978-3-031-02428-3

A Publication in the Springer series
SYNTHESIS LECTURES ON MATHEMATICS AND STATISTICS

Lecture #36
Series Editor: Steven G. Krantz, *Washington University, St. Louis*
Series ISSN
Print 1938-1743   Electronic 1938-1751

Crowd Dynamics
by Kinetic Theory Modeling

Complexity, Modeling, Simulations, and Safety

Bouchra Aylaj
Université Hassan II de Casablanca

Nicola Bellomo
University of Granada

Livio Gibelli
University of Edinburgh

Damián Knopoff
Centro de Investigación y Estudios de Matemática (CONICET) and Famaf (UNC)

SYNTHESIS LECTURES ON MATHEMATICS AND STATISTICS #36

ABSTRACT

The contents of this brief Lecture Note are devoted to modeling, simulations, and applications with the aim of proposing a unified multiscale approach accounting for the physics and the psychology of people in crowds. The modeling approach is based on the mathematical theory of active particles, with the goal of contributing to safety problems of interest for the well-being of our society, for instance, by supporting crisis management in critical situations such as sudden evacuation dynamics induced through complex venues by incidents.

KEYWORDS

crowd modeling, classical kinetic theory, active particles models, scaling, mathematical structures

Contents

Preface

The modeling, qualitative, and computational analysis of human crowds is an interdisciplinary research field which involves a variety of challenging analytic and numerical problems, generated by the derivation of models followed by their application to real-world dynamics. The overall approach requires not only tools of physics and mathematics, but also additional knowledge concerning social and psychological behaviors of human crowds.

The study of crowd dynamics can contribute to tackle safety problems of interest for the well-being of our society, for instance, by supporting crisis management in critical situations such as sudden evacuation dynamics induced through complex venues by incidents. Simulations can support crisis managers to handle the competition of antagonist groups in a crowd.

Further possible applications are met in the field of engineering sciences, for instance the design of internal venues in building can be optimized in order to reduce the evacuation time, or in the design of structures, as lively bridges, where crowd simulations contribute to estimate the loads applied by the crowd to the bridge.

This Lecture Note aims at tackling the conceptual difficulties generated by the fact that the physics, needed by the modeling and simulation of complex (living) systems, is not fully understood, while mathematical models, at the present state of the art, do not yet seem able to depict the whole variety of emerging behaviors which are observed in crowd dynamics. In addition, individual behaviors and heterogeneity can lead, for instance in crisis circumstances, to large deviations with respect to the usual dynamics in rational flow conditions. These irrational behaviors might generate situations of high risks for humans involved in a chaotic "irrational" crowd.

The development of the modeling approach proposed in this Lecture Note accounts also for the influence of emotional states of the crowd on the overall dynamics, thus investigating emergent behaviors and pattern formations.

The content is primarily focused on modeling, simulations, and applications with the aim of proposing a unified multiscale approach to the mathematical interpretation of the physics and the psychology of crowds. Analytic problems, typical of mathematical sciences, are brought to the attention of the interested reader, but are not specifically treated here.

The complexity features are already introduced in Chapter 1 which presents a brief introduction to crowd modeling and motivations to develop a research activity in the field in view of applications. Subsequently, this chapter presents the plan of the Lectures Note.

Chapter 2 provides a deeper analysis of the complexity features of human crowds and defines, accordingly, a modeling strategy.

Chapter 3 presents the theoretical tools to be used toward modeling from the classical kinetic theory to the statistical dynamics of active particles.

Chapter 4 selects a model with the ability of accounting for the geometry and the quality of the venues, as well as of the emotional state pervading homogeneously the crowd. Various simulations investigate the predictive ability of the model.

Chapter 5 indicates some possible research perspectives which are followed by hints to tackle them according to the aim of building a bridge between the present state of the art and future achievements. In more detail, the following topics are taken into account: further studies on the role of social dynamics in crowds; a multiscale vision of crowd dynamics; and support by mathematical models and artificial intelligence to crisis management. Last, a Closure, inserted in Chapter 5, presents some perspective ideas which outline, according to the authors' bias, how advanced research activity will/should be in the next decade.

Bouchra Aylaj, Nicola Bellomo, Livio Gibelli, and Damián Knopoff
October 2020

CHAPTER 1

Complexity of Human Crowds and Modeling Strategy

Abstract: This chapter provides a presentation of the motivations, methodological approach, and overall contents of this Lecture Note devoted to modeling, simulations, and safety problems of human crowds. The modeling accounts for the behavioral, social, and emotional well-being of individuals in crowds. A key topic in this chapter is the identification of the specific features of human crowds to be viewed as a living, hence complex, system. The modeling strategy consists of looking for a rationale toward the derivation of mathematical structures suitable to capture these complexity features. An introduction to a strategy to validate models, hence to support their application, is introduced in view of more extended space focused on the specific models proposed in the next chapters.

1.1 INTRODUCTION

This Lecture Note provides a survey and critical analysis of the existing literature on human crowd dynamics with a focus on modeling, validation, applications, and simulations. Pursuing this objective requires, first, an understanding of the complex features of human crowds accounting for individual and collective psychological behaviors of individuals in the crowd [30], and subsequently, tackling a variety of highly challenging analytic and computational problems that the physics and the complexity features of human crowds pose to applied mathematicians and physicists involved in the modeling approach.

For more details on methodological issues, we mention that the approach is based on the so-called kinetic theory for active particles [4] which we have selected toward the modeling approach after some technical comparisons with the other modeling scales, namely microscopic (individual-based) and macroscopic (hydrodynamical). Indeed, this choice appears, based on the authors bias, appropriate for capturing the main features of crowds to be viewed as a living, hence complex, system. The idea that we apply is that modeling is required to capture, at the greatest possible extent, the complexity features of crowds.

An important objective of the Lecture Note consists in showing how modeling and computational simulations can contribute to tackle safety problems of interest for the well-being of our society. For instance, by addressing research activity toward crisis management of human crowds in critical situations such as evacuation dynamics induced by incidents or competition of antagonist groups in a crowd. This topic is treated, as we shall see, by an introduction to the use of some tools of artificial intelligence.

This first chapter is devoted to a general introduction to the overall contents of the Lecture Note starting from an introduction to a modeling strategy which pervades all following chapters. The contents are as follows. Section 1.2 starts from a preliminary survey of the motivations leading to the study of human crowds and subsequently reports about a selection of research articles and books chosen out of the pertinent literature. Section 1.3 brings to the reader's attention a selection of complexity features of human crowds according to the aim of providing the framework of the specific features which should be accounted for in the modeling approach. Section 1.4 provides some preliminary ideas toward a possible strategy to derive and validate crowd models. Section 1.5 provides a description of the overall plan of the Lecture Note.

1.2   MOTIVATIONS TOWARD CROWD MODELING AND SIMULATIONS

The modeling, qualitative, and computational analysis of human crowds, as mentioned in the Preface, can be viewed as an interdisciplinary research field which generates a variety of challenging analytic and numerical problems related to the derivation of models as well as by their application to real-world dynamics. The approach to modeling and simulations is required to go beyond scientific motivations and to look at the potential benefits for the society. As an example, the realistic modeling of human crowds can lead to simulations which can support crisis managers to handle emergency situations, such as sudden and rapid evacuations through complex venues.

The main conceptual difficulty is that the physics, needed by the modeling and simulation of complex (living) systems, in general, is not fully understood. Indeed, mathematical models, at the present state of the art, do not yet seem able to depict the whole variety of their emerging behaviors. The main difficulty is that, unlike inert matter, causality principles are not known in the case of living systems, while, whenever available, causes and effects cannot be mapped linearly. A consequence of this type of nonlinearity is that similar causes might have different effects, which is a large deviation effect, and even that different causes might generate similar effects.

In addition, individual behaviors and heterogeneity, in the case of crisis circumstances, can lead to large deviations with respect to the usual dynamics in rational flow conditions. These irrational behaviors might generate situations of high risks for a crowd which might become "chaotic" and "irrational." An interesting dynamics, to be taken into account in the modeling approach, appears to be the role of stress contagion over the movement of the crowd, where stress induced by localized overcrowding propagate over the whole crowd and consequently affect the safety of the people [21, 23, 29, 32].

The existing literature on the mathematical modeling of human crowds is reported in some survey papers, which offer to applied mathematicians different view points and modeling strategies in a field, where a unified, commonly shared, approach does not exist yet. A preliminary selection of research and survey papers, as well as books, can be brought to the attention of the interested readers to enlighten all the above reasonings.

- The review [17] presents and critically analyzes the main features of the physics of a crowd viewed as a multi-particle system and focuses on the modeling at the microscopic scale for pedestrians undergoing individual-based interactions. A vision of multi-particle systems, whose dynamics can be modeled by tools of statistical physics, is also presented.

- The survey [20] and the book [13] deal with the modeling at the macroscopic scale, by methods analogous to those of hydrodynamics, where one of the most challenging conceptual difficulties consists of understanding how the crowd, viewed as a continuum, selects the velocity direction and speed by which pedestrians move. See also [28] on the mathematical theory of time-evolving measures in the macroscopic modeling of pedestrian flow.

- The research articles [5, 7] have proposed the concept of the crowds as a living, hence complex, system. This approach requires the search for mathematical tools suitable to take into account, as far as it is possible, the complexity features of the system under consideration. Contributions to model rational and emotional behaviors of crowds have been given by various authors [2, 8, 9, 15, 31]

- Scaling problems and related mathematical topics are treated in the book [13], while [3] shows how models at the macroscopic (hydrodynamical) scale can be derived from the underlying description delivered at the mesoscopic scale. A hierarchy of heuristic models is proposed in [14]. Coupling the dynamics of vehicles and pedestrians has been studied in [10].

- The support of mathematical–computational models to crisis management during evacuation is critically analyzed in the survey [6] based on the study of a broad literature [12, 18, 19, 23–25, 29, 32] mainly developed in the last decade under the need of supporting safety managers.

The various chapters of the edited book [16], which has been recently published, present various contributions on the aforementioned topics including the achievement of empirical data and control problems, while a critical analysis of the state of the art indicates that the following issue has not yet been exhaustively treated.

Our view point is also expressed by the following remark from [8]:

The greatest part of known models are based on the assumption of rational behaviors of individuals. However, real conditions can show a presence of irrational behaviors that can generate events where safety conditions are damaged. When these conditions appear, small deviations in the input create large deviations in the output.

The literature on the practical management of real crowd dynamics problems, including crisis and safety problems, have enlightened that social phenomena pervade heterogeneous crowds and can have an important influence on the interaction rules. Therefore, both social and mechanical dynamics, as well as their complex interactions, should be taken into account.

As it is known, the modeling approach can be developed at the three usual scales, namely *microscopic, macroscopic*, and the intermediate *mesoscopic scale*, the latter is occasionally called *kinetic*. However, none of them is fully satisfactory. In fact, accounting for the modeling of multiple interactions, as well as for the heterogeneous behavior of the crowd, is not immediate in the case of various known models at the microscopic scale. Macroscopic models do not account for the aforementioned heterogeneity unless mixture theories are developed.

Although we are well aware of the aforementioned drawbacks, the selection of the most appropriate scale is the first step toward the modeling of human crowds. Accordingly, the kinetic theory approach appears, according to our bias, to be more flexible as it can overcome, at least partially, the previously mentioned drawbacks. However, additional work is needed to develop them toward the challenging objectives treated in the following chapters, e.g., dealing with the modeling of the complex interaction between social and mechanical dynamics, but also deriving models suitable to reproduce the dynamics of the system at different densities from the low values to the high ones.

Some introductory concepts have been proposed in the literature by the kinetic theory approach, starting from [3, 5], where, for example, an additional parameter has been inserted in the microscopic state of individual entities to describe stress conditions. More recently, [31] considers a dynamics in one space dimension described at the macroscopic scale, where panic is propagated by a BGK-type model [11, 22]. The propagation in space of emotional states has been studied in [8]. A study on the role of social dynamics on individual interactions with influence at the higher scale is developed in [14, 15].

An important motivation is given by the strategic objective of designing mathematical models suitable to depict the complexity features of the crowd and to reproduce, as far as it is possible, empirical data. Simulations are required to reproduce real flow conditions and can allow support crisis managers to act in a way to improve safety conditions. It is not an easy task, as the modeling approach cannot be limited to account for mechanical rules, as human behaviors have to be included in it as these heterogeneous behaviors can substantially modify the walking strategy.

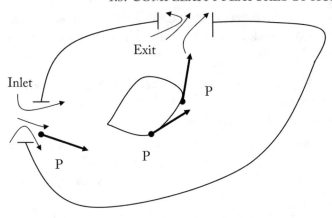

Figure 1.1: Domain with inlet, exit, and internal obstacles.

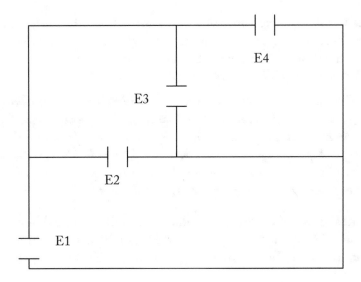

Figure 1.2: A sequence of interconnected areas.

1.3   COMPLEXITY FEATURES OF HUMAN CROWDS

Let us consider the dynamics of human crowds in venues which include inlet and outlet doors, as well as internal obstacles. The dynamics of a crowd often occur over complex geometries, where individuals move across areas characterized by different geometrical and physical properties.

An example is shown in Fig. 1.1, where the set of all walls, including that of obstacles and of entrance and exit doors, is denoted by Σ, while Fig. 1.2 shows the geometry of a possible sequence of walking areas from the inlet E_1 to the outlet E_4 through the intermediate passages E_2 and E_3.

Some definitions to characterize the specific features of a human crowd have been given in [19], these definitions are reported in the following, implemented where some preliminary speculations follow in italic.

- **Definition of crowd:** Agglomeration of many people in the same area at the same time. The density of people is assumed to be high enough to cause continuous interactions, or reactions, with other individuals.

  This definition clearly states that one can talk about a crowd if the number of interactions per unit time is significant. An important perspective would be looking for a characteristic number suitable to identify the transition from individual and collective dynamics. In addition, one should study how such a "possible" number can be related to the selection of the modeling scale selecting between microscopic (individual based), mesoscopic (kinetic), and macroscopic (hydrodynamical).

- **Collective intelligence:** Emergent functional behavior of a large number of people that results from interactions of individuals rather than from individual reasoning or global optimization. Establishment of a qualitatively new behavior through nonlinear interactions of many individuals.

  Each individual has a walking strategy by which he/she selects a walking direction and a speed, e.g., to walk out from an area with walls, obstacles, and exits as shown in Figs. 1.1 and 1.2. However, this strategy is subject to an influence coming from the whole crowd basically induced by two specific engines, namely "mechanical," as individual look for less crowded paths to improve their speed, and "psychological," as individual are somehow attracted by other individuals, thus leading to different emergent behaviors.

- **Panic breakdown of ordered, cooperative behavior of individuals:** Often, panic is characterized by an attempted escape of many individuals from a real or perceived threat in situations of a perceived struggle for survival, which may end up in trampling or crushing.

  The key problem consists in understanding how these stress conditions are generated and propagated in space due to communications, visual and vocal, between individual entities. It is well understood that the level of emotional state modifies the walking strategy.

These definitions and the comments, which have followed each definition, can be viewed as the first step toward a necessary deep understanding of the complexity features of human crowds. Indeed, the recent literature has enlightened the need of a modeling approach, where the behavioral features of crowds, to be viewed as a living, complex system, should be taken into account.

The most important feature appears to be the ability to express a strategy which is heterogeneously distributed among walkers and depends on their own state and on that of the entities in their surrounding walkers and environment. Heterogeneity can include, as an example, a possible presence of leaders, who attract the crowd to their own strategy. For instance, leaders can contribute to the selection of optimal routes among the various available ones. An additional feature to be accounted for is the possible presence of antagonist groups with possible dynamics across groups, where groups not only contrast each other, but the dynamics might even include transition from one group to the other.

Both strategy and interactions are influenced by stress conditions which, in some cases, are simply induced by overcrowding. Stress can have an important influence on the dynamics crowds and can affect the dynamics up to safety conditions. In fact, stress conditions induce an excess of aggregation as all individuals rush toward the same direction and attempt to increase their speed even in high density conditions.

Therefore the modeling approach requires a unified vision of the *complexity features* of human crowds to be taken into account in the modeling strategy. A useful reference is given in paper [7] which is here revisited in five items. The reader can rapidly understand how these complexity features are somehow related to the above definitions.

- **Ability to express a strategy:** Living entities are capable to develop specific *strategies* and *organization abilities* which depend on the state of the surrounding environment. These can be expressed even in absence of any external *organizing principle*. Namely, the strategy belongs to the walking individual based on her/his previous learning experience and it might be modified by interactions with other individuals.

- **Learning ability:** Living systems receive inputs from their environments and have the *ability to learn from past experience*. Therefore, their strategic ability and the rules of interactions evolve in time due to communications linking each individual to the others. Learning dynamics can last a short time, but it can also leave a permanent sign in human minds. Individuals can learn also from external inputs such as those coming from external signaling and/or vocal messages.

- **Heterogeneity:** The ability to express a strategy is not the same for all walkers: *Heterogeneity* can include, in addition to different walking abilities and strategies, different objectives, a possible presence of leaders, as well as the presence of antagonist groups.

- **Nonlinear Interactions:** Interactions are nonlinearly additive and involve immediate neighbors, but in some cases also distant particles due to their ability to communicate. The concept of nonlinearity means that the output of interactions can be influenced by the dependent variable. For instance, local density has an influence on the stress which, in turn, has an influence on the strategy and walking dynamics.

- **Influence of environmental conditions:** The dynamics is affected by the quality of environment, weather conditions, and geometry of the venue. In general, the quality of the venue where walkers move has an influence over the speed, namely a high-quality walking area and good weather conditions support the trend toward high speed. However, this is not simply a matter of quality, but also of geometry as certain geometries induce regular flows in contrast with high-density flows, namely with an important influence on the overall dynamics.

All features reported in this section can have an important influence on the collective dynamics whatever is the representation and scale selected for the modeling approach. Indeed, this aspect, according to our bias, is the key feature of the research topic treated in this lecture. In addition, modeling and simulations should take into account that the dynamics depend on quality features of the area, where the crowd moves, as well as on the geometry of the venue which can be constituted by several interconnected areas and might include internal obstacles. In general, each area might be characterized by different quality features.

1.4   MODELING STRATEGY AND VALIDATION

The strategic modeling objective consists in designing mathematical models suitable to depict the complexity features of the crowd which have been stated in the preceding section. In addition, models are required to reproduce, as far as it is possible, empirical data. In general, the modeling approach should account for the specific use of models in the application. Some examples of possible finalizations of computational models are given without claiming to be exhaustive.

1. Support crisis managers to take rapid decisions in critical situations, for instance evacuations in the case of incidents, but, in some cases, simply to expedite the flow of pedestrians.

2. Training crisis managers to handle a broad variety of possible scenarios of crowd dynamics in complex domains under a possible variety of external actions to support safety conditions which require that local density remains below a safety level for a dynamic which depend on the shape and quality of venue.

3. Development of test simulations to investigate the role of the geometry and quality of venues to support safety conditions by an appropriate design of the geometry of venues, hence simulations can be developed to optimize the design of the areas where crowds move.

4. Modeling the contrast between security forces and antagonists, as well as between different antagonist groups in a public demonstration.

5. Engineering applications where computational models can contribute to improve the geometry of buildings or to estimate loads in crowd–structure interactions.

The sample applications which have been mentioned in the above items are mainly focused on safety and security problems. Indeed, the need of developing a research activity in this specific field is strongly motivated, as already mentioned, by the literature in journals of inter-disciplinary engineering sciences, where the dynamics occur. However, it is worth mentioning that, in addition to them, applied mathematicians and physicists are attracted by the highly chal-lenging analytic and computational problems. Therefore, this Lecture Note aims also at showing how the study of crowd dynamics can contribute to the scientific development of mathematical sciences.

Bearing all the above in mind, let us indicate the sequential steps of a strategy toward derivation of models and their simulations toward specific applications of models.

1 – Derivation of a mathematical structure to support the modeling approach: The modeling approach refers, as we shall see in the next chapter, to the strategic selection of one of the three classical scales, namely microscopic (individua-based), mesoscopic (kinetic), and macroscopic (hydrodynamic). The selection of the scale should be developed by a preliminary analysis of the possible ability of models to reproduce, as far as it is possible, the complexity features of living systems.

Mathematical models should be derived within this specific framework, while validation, computational problems, and the ability of models to depict the dynamics in complex venues follow only if this preliminary step has been accomplished. This structure offers the background for the derivation of models based on a detailed analysis and modeling of walkers (pedestrians) among themselves. In addition, the quality and main physical features of the venue, where the dynamics occur, should be taken into account. For example, venues which include inlet and outlet doors and the presence of internal obstacles.

2 – Modeling the dynamics of emotional states: Simulations should be developed to account all aspects of the complex heterogeneity of crowds, where emotional states are transferred by contagion related to vocal and/or visual signaling. Modeling should account both for intensity and localization in space of the said states which propagate by visual and vocal communications. Stress can induce irrational behaviors.

3 – Validation of models: Validation of models should be based on their ability to reproduce, quantitatively, available empirical data, and qualitatively expected emerging behaviors. The for-mer specifically refers to the so-called velocity diagrams (and fundamental diagrams), namely mean velocity vs. density (and flow vs. density), where one of the difficulties is due to the fact that the diagram provides a macroscopic information, while models might be at the micro-scopic scale. The latter is based on the empirical observation that collective motions exhibit a self-organization ability leading to patterns which are reproduced qualitatively, but it might be subject to large deviations for small variations of the flow conditions.

4 – Derivation of a computational code: The design of a computational code is related to the scale, selected toward modeling, as different numerical tools are used for each scale. Namely, time integration methods for large systems of ordinary differential equations, particle methods in the case of the kinetic theory approach, and finite differences or volumes in the case of the

Figure 1.3: Continuum flow patterns.

macroscopic approach. The requirements for simulations basically consists of the achievements of a certain accuracy in computing local density and velocity and of computational times lower or equal to the real time of the crowd.

5 – Databases and big data managing: The possible variety of real flow situations can be interpreted by simulations to be stored in a database to support crisis managers who have the task of inducing situations of minor overcrowding. A strategy might be developed to use the database as a predictive engine to select optimal actions to support crisis. It is worth stressing that simulations can contribute to crisis management not only by showing that the model can provide an accurate description of the crowd dynamics, but also by verifying how the evacuation time increases under stress conditions and the identification of a risk situation due to an excessive concentration of walkers in the same area. Indeed, safety conditions require that local density remains below a safety level. These dynamics depend on venue parameters, hence, simulations can contribute to optimize the areas of the venue for crowds.

Let us discuss further the strategic selection of one of the three classical scales which will be technically treated in the next chapter. Looking at Fig. 1.3, the pattern occupied by the crowd appears to be continuous, which might provide the argument that the hydrodynamical approach is the most appropriate, also considering that it requires a lower computational time. On the other hand, Fig. 1.4 shows the presence of both continuous and rarefied patterns which supports the use of the kinetic theory approach, while Fig. 1.5 shows a rarefied presence of walkers, thus invoking individual-based models.

Figure 1.4: Rarefied and overcrowded patterns in the same venue.

Figure 1.5: Individual-based dynamics.

However, anticipating the contents of Chapter 2, we need to stress that the selection of the scale is a tricky problem which cannot be rapidly (and naively) treated as we have done here. In fact, the flow can be characterized by a contextual presence of highly rarefied and continuous zones. In addition, flow patterns are highly influenced by the geometry of the areas where the crowd moves as well as by the emotional state. Therefore, we have simply anticipated a key problem which deserves attention and further deep analysis.

1.5   PLAN OF THE LECTURE NOTES

The introduction delivered in the preceding sections opens our mind to provide a description of the next chapters.

Chapter 2 presents the general mathematical structures which offer the conceptual framework for the derivation of models. The presentation refers to the three scales, namely microscopic (individual-based), mesoscopic (by kinetic theory tools), and macroscopic (hydrodynamical). Structures both for first-order and second-order models are given thus explaining the conceptual differences between the two classes of structures. Subsequently, the contents of the chapter explains how mathematical model can, and should, include models on the role of emotional states and how these states can propagate in the crowd by a sort of contagion related to communications by vocal and visual signals and hence by a collective learning dynamics, taking into account for the quality of the environment and the geometry of the venue. Finally, a critical analysis is proposed to enlighten how the choice of the kinetic theory approach deserves to be selected toward the modeling approach.

Chapter 3 is devoted to technical aspects of the mathematical tools of the kinetic theory starting from a concise presentation of the classical Boltzmann and Vlasov equations which are the important models that can contribute to the derivation of a mathematical kinetic theory of crowd dynamics. This chapter enlightens how classical models of the mathematical kinetic theory cannot be straightforwardly applied to model human crowds. Indeed, substantial developments, mainly focused on the modeling of interactions, are necessary to account for the specific behavioral features of human crowds. Accordingly, a general structure is derived to provide the conceptual basis toward the design of specific models accounting also for the aforementioned propagation of emotional states.

Chapter 4 provides a review of models, from [5]–[8], derived by the kinetic theory approach as they have been developed in the last decade somehow following the analogous, however presenting important differences, approach to the modeling of vehicular traffic initiated by the pioneer articles delivered by Prigogine and Hermannn [27] further developed by Paveri Fontana [26]. Out of this general review, a well-defined model is selected, while simulations follow to enlighten its predictive ability to depict emerging behaviors in view of the validation of models.

Chapter 5 is devoted to some research perspectives, selected according to the authors' bias, which are brought to the attention of the interested reader toward possible research programs. The following topics have been selected: a multiscale vision of crowd dynamics; a systems approach to crowd dynamics; and reasonings on safety and security research activity which can be developed also by tools of artificial intelligence to support management of crisis situations. This section ends with a closure presenting some free speculations on the research perspectives in the next decade.

1.6   BIBLIOGRAPHY

[1] B. Aylaj, N. Bellomo, L. Gibelli, and A. Reali, On a unified multiscale vision of behavioral crowds, *Mathematical Models and Methods in Applied Sciences*, 30(1):1–22, 2020. DOI: 10.1142/S0218202520500013

[2] R. Bailo, J. A. Carrillo, and P. Degond, Pedestrian models based on rational behaviors, *Crowd Dynamics, Volume 1: Theory, Models, and Safety Problems*, Birkhäuser-Springer, 2018. DOI: 10.1007/978-3-030-05129-7_9 3

[3] N. Bellomo and A. Bellouquid, On multiscale models of pedestrian crowds from mesoscopic to macroscopic, *Communications in Mathematical Sciences*, 13(7):1649–1664, 2015. DOI: 10.4310/cms.2015.v13.n7.a1 3, 4

[4] N. Bellomo, A. Bellouquid, L. Gibelli, and N. Outada, *A Quest Towards a Mathematical Theory of Living Systems*, Birkhäuser-Springer, New York, 2017. DOI: 10.1007/978-3-319-57436-3 1

[5] N. Bellomo, A. Bellouquid, and D. Knopoff, From the micro-scale to collective crowd dynamics, *Multiscale Modelling and Simulation*, 11:943–963, 2013. DOI: 10.1137/130904569 3, 4, 12

[6] N. Bellomo, D. Clark, L. Gibelli, P. Townsend, and B. J. Vreugdenhil, Human behaviours in evacuation crowd dynamics: From modelling to "big data" toward crisis management, *Physics of Life Reviews*, 18:1–21, 2016. DOI: 10.1016/j.plrev.2016.05.014 3

[7] N. Bellomo and L. Gibelli, Toward a behavioral-social dynamics of pedestrian crowds, *Mathematical Models and Methods in Applied Sciences*, 25:2417–2437, 2015. DOI: 10.1142/S0218202515400138 3, 7

[8] N. Bellomo, L. Gibelli, and N. Outada, On the interplay between behavioral dynamics and social interactions in human crowds, *Kinetic and Related Models*, 12:397–409, 2019. DOI: 10.3934/krm.2019017 3, 4, 12

[9] A. L. Bertozzi, J. Rosado, M. B. Short, and L. Wang, Contagion shocks in one dimension, *Journal Statistical Physics*, 158(3):647–664, 2015. DOI: 10.1007/s10955-014-1019-6 3

[10] R. Borsche, A. Klar, S. Köhn, and A. Meurer, Coupling traffic flow networks to pedestrian motion, *Mathematical Models and Methods in Applied Sciences*, 24:359–380, 2014. DOI: 10.1142/s0218202513400113 3

[11] C. Cercignani, R. Illner, and M. Pulvirenti, *The Mathematical Theory of Diluted Gas*, Springer, Heidelberg, New York, 1993. DOI: 10.1007/978-1-4419-8524-8 4

[12] A. Corbetta, A. Mountean, and K. Vafayi, Parameter estimation of social forces in pedestrian dynamics models via probabilistic method, *Mathematical Biosciences Engineering*, 12:337–356, 2015. DOI: 10.3934/mbe.2015.12.337 3

[13] E. Cristiani, B. Piccoli, and A. Tosin, *Multiscale Modeling of Pedestrian Dynamics*, Springer Italy, 2014. DOI: 10.1007/978-3-319-06620-2 3

[14] P. Degond, C. Appert-Rolland, M. Moussaïd, J. Pettré, and G. Theraulaz, A hierarchy of heuristic-based models of crowd dynamics, *Journal Statistical Physics*, 152:1033–1068, 2013. DOI: 10.1007/s10955-013-0805-x 3, 4

[15] P. Degond, J.-G. Liu, S. Merino-Aceituno, and T. Tardiveau, Continuum dynamics of the intention field under weakly cohesive social interaction, *Mathematical Models and Methods in Applied Sciences*, 27:159–182, 2017. DOI: 10.1142/s021820251740005x 3, 4

[16] L. Gibelli and N. Bellomo, *Crowd Dynamics, Volume 1: Theory, Models, and Safety Problems*, Birkhäuser-Springer, 2018. DOI: 10.1007/978-3-030-05129-7 3

[17] D. Helbing, Traffic and related self-driven many-particle systems, *Review Modern Physics*, 73:1067–1141, 2001. DOI: 10.1103/revmodphys.73.1067 3

[18] D. Helbing, I. Farkas, and T. Vicsek, Simulating dynamical feature of escape panic, *Nature*, 407:487–490, 2000. DOI: 10.1038/35035023 3

[19] D. Helbing and A. Johansson, Pedestrian crowd and evacuation dynamics, *Encyclopedia of Complexity and System Science*, pages 6476–6495, Springer, 2009. DOI: 10.1007/978-3-642-27737-5_382-5 3, 6

[20] R. L. Hughes, The flow of human crowds, *Annual Review Fluid Mechanics*, 35:169–182, 2003. DOI: 10.1016/s0378-4754(00)00228-7 3

[21] M. Kinateder et al., Human behaviour in severe tunnel accidents: Effects of information and behavioural training, *Transportation Research Part F: Traffic Psychology and Behaviour*, 17:20–32, 2013. DOI: 10.1016/j.trf.2012.09.001 2

[22] M. Kogan, *Rarefied Gas Dynamics*, Plenum Press, New York, 1968. DOI: 10.1007/978-1-4899-6381-9 4

[23] J. Lin and T. A. Luckas, A particle swarm optimization model of emergency airplane evacuation with emotion, *Networks Heterogeneous Media*, 10:631–646, 2015. DOI: 10.3934/nhm.2015.10.631 2, 3

[24] M. Moussaid, D. Helbing, S. Garnier, A. Johanson, M. Combe, and G. Theraulaz, Experimental study of the behavioral underlying mechanism underlying self-organization in human crowd, *Proc. Royal Society B: Biological Sciences*, 276:2755–2762, 2009. 3

[25] M. Moussaïd and G. Theraulaz, Comment les piétons marchent dans la foule, *La Recherche*, 450:56–59, 2011. 3

[26] S. Paveri Fontana, On Boltzmann like treatments for traffic flow, *Transportation Research*, 9:225–235, 1975. DOI: 10.1016/0041-1647(75)90063-5 12

[27] I. Prigogine and R. Herman, *Kinetic Theory of Vehicular Traffic*, Elsevier, New York, 1971. 12

[28] B. Piccoli and A. Tosin, Time-evolving measures and macroscopic modeling of pedestrian flow, *Archives Rational Mechanics Analysis*, 199:707–738, 2011. DOI: 10.1007/s00205-010-0366-y 3

[29] F. Ronchi, F. Nieto Uriz, X. Criel, and P. Reilly, Modelling large-scale evacuation of music festival, *Fire Safety*, 5:11–19, 2016. DOI: 10.1016/j.csfs.2015.12.002 2, 3

[30] A. Templeton, J. Drury, and A. Philippides, From mindless masses to small groups: Conceptualizing collective behavious in crowd modeling, *Review General Psychology*, 19:215–229, 2015. DOI: 10.1037/gpr0000032 1

[31] L. Wang, M. Short, and A. L. Bertozzi, Efficient numerical methods for multiscale crowd dynamics with emotional contagion, *Mathematical Models and Methods in Applied Sciences*, 27:205–230, 2017. DOI: 10.1142/s0218202517400073 3, 4

[32] N. Wijermans, C. Conrado, M. van Steen, C. Martella, and J. L. Li, A landscape of crowd management support: An integrative approach, *Safety Science*, 86:142–164, 2016. DOI: 10.1016/j.ssci.2016.02.027 2, 3

CHAPTER 2

Scaling and Mathematical Structures

Abstract   This chapter shows how a strategy to derive and validate models can be developed for crowd dynamics which can occur in complex venues constituted by several interconnected areas, each of them presenting different geometrical and physical features. The most important feature is the ability of individuals to express a walking strategy which is heterogeneously distributed among walkers and depends on their own mechanical and emotional state, on that of the entities in their surrounding walkers, as well as on the geometry and quality of the areas where the walkers move. An additional feature of the strategy is the selection of the modeling scale, which corresponds to different ways of representing the system and the related dynamics. The following representations are examined: individual, statistical, and continuum, while a critical analysis leads to the selection of the kinetic theory approach.

2.1   INTRODUCTION

This chapter provides an introduction to a strategy to derive and validate models deemed to describe flow conditions of human crowds in venues which include inlet-outlet doors, surrounding walls, and internal obstacles. Dynamics can occur in complex venues constituted by several interconnected areas, each of them presenting different geometrical and physical features. The recent literature on crowd modeling has enlightened the need of a modeling approach, where the behavioral features of crowds, to be viewed as a living, hence complex system, are taken into account. The most important feature appears to be the ability to express a walking strategy which is heterogeneously distributed among walkers and depends on their own mechanical and emotional state, on that of the entities in their surrounding walkers, as well as on the geometry and quality of the areas where the walkers move.

The meaning of heterogeneity specifically means that walking ability is not the same for all individuals, but it spans from low values for individuals with limited walking ability to high values for individuals with athletic ability. In addition, the walking strategy also is not developed in the same way by all walkers, as an example it depends on the level of stress which differs across individuals. Heterogeneity in the crowd also can include groups of individuals with different purposes, for instance it can include a possible presence of leaders, who attract the crowd to their own strategy suitable to select optimal trajectories among the various available ones, or simply pedestrians which move toward different exits or meeting points.

An additional feature, to be taken into account, is the emotional state, for instance stress conditions which, in some cases, are simply induced by overcrowding. This feature is heterogeneously distributed and can have an important influence on the dynamics ultimately affecting safety. Interesting empirical studies have been developed on this topic focusing precisely on the influence of stress conditions and limited visibility due to geometrical and quality properties of the areas, where the crowd moves [26, 27].

Recent events indicate that the modeling approach should also consider crowds where social contrasts can appear, as an example between rioters and security forces. Indeed, this is a highly challenging objective for scientists involved on crowd modeling related to applications which can have an impact on society.

The selection of the modeling scale, which corresponds to different ways of representing the system and the related dynamics, is a key initial step in the modeling approach. In general, the following representations can be selected: individual, statistical, and continuum. Within each scale, different computational approximation methods can be developed, either mathematical, for instance finite differences of finite volumes, or physical, for instance lattice methods.

The derivation of a model is not, however, the end of the story as it is necessary to prove its validity, based on an appropriate use if empirical data, focusing on the specific applications under consideration. Subsequently, applications require the development of computational codes which have to be technically related to the scale selected for the modeling approach.

The various topics, which have been briefly introduced in this section, suggest the following plan of the chapter. Section 2.2 introduces the reader to general topics of crowd modeling by defining the modeling scales and to the mathematical structures deemed to offer the conceptual basis, at each scale, for the derivation of models. A selection of scientific contributions shows how the literature developed modeling and simulations approaches at each scale. Section 2.3 presents a modeling strategy which consists in showing how models can be derived by implementing the said structures with heuristic models suitable to describe, at each scale, interactions involving individuals in the dynamics. This section also introduces some preliminary concepts on the validation problem and indicates how empirical data can be used to estimate the reliability of models in general and specifically related to possible applications. Section 2.4 motivates the selection of the mesoscopic scale, namely of the kinetic theory approach, as the most appropriate to capture, by mathematical equations, the complexity features of human crowds. Section 2.5 finally proposes a critical analysis on the contents of the chapter looking ahead to research perspectives.

2.2   MATHEMATICAL STRUCTURES

This section provides a description of the representation of human crowds at the three classical scales, namely microscopic, mesoscopic, and macroscopic which will be defined precisely in this section. The contents technically refer to [1, 8], where a multiscale vision is proposed with the aim of showing how the modeling approach should be developed within a multiscale framework accounting for different mathematical representations and related mathematical structures. Let us consider the crowd dynamics in a two-dimensional domain which includes, as mentioned, inlet and outlet doors as well as internal obstacles. We denote by Σ the overall geometry of the domain including the localization of inlets and outlets as well as the shapes of the walls and obstacles as it has been represented in Chapter 1, while $\Sigma_0 \subseteq \Sigma$ denotes the region which contains the whole crowd at the initial time $t = t_0$.

The following reference physical quantities and variables are introduced in view of the definition of the aforementioned frameworks.

- ℓ: Characteristic length to be taken as the diameter of the circle containing the domain where the crowd moves, while if the crowd is located in unbounded domains, it refers to the diameter of the circle containing the domain Σ_0 of the initial displacement of the crowd.

- v_M: Highest individual speed which can be reached by a very fast walker in a free flow, namely a low-density flow, in high-quality venues.

- $v_\ell \leq v_M$: Highest individual, venue dependent, speed in a free flow.

- $T = \ell/v_M$: Characteristic time corresponding to the time needed by a fast walker to cover the distance ℓ in a high-quality venue.

- ρ_M: Maximum number density (occupancy) of walkers packed in a square meter.

- Ω: Visibility domain, to be defined precisely at each scale, depending on the position of the walker and on the overall geometry of the venue where the walkers move.

In addition, a minimal model should include, at least, the following parameters which refer to two key features which have a significant impact on the dynamics of walkers in a crowd.

- $\alpha \in [0, 1]$: Parameter modeling the overall quality of the venue, where $\alpha = 0$ corresponds to very low quality which prevents motion, while $\alpha = 1$ to very high quality allowing fast motion. This parameter depends on the position of walkers according to the quality of the areas where they move.

- $\beta \in [0, 1]$: Parameter modeling the mean stress of the crowd, where $\beta = 0$ and $\beta = 1$ correspond, respectively, to very low and high limit stress.

Dimensionless quantities are used referring them to the unit size domain $[0, 1]$, where linear space variables are referred to ℓ, real time to T, and speed is referred to v_M. In addition, macroscopic quantities, specifically density and mean speed, are divided by ρ_M and v_M. Hence, their physical meaning is below one. The following simple model $v_\ell = \alpha\, v_M$ has been proposed [7] to refer v_ℓ to $\mathbf{v}_M$. Namely, the venue dependent maximal speed v_ℓ is reduced by α with respect to the maximal speed $\mathbf{v}_M$ in optimal quality venues.

Figure 2.1: Overcrowded patterns.

The specific emotional state selected in the modeling approach is the stress which is accounted for by means of the parameter β. The role of this parameter on the motion is defined later focusing on interactions. It is worth mentioning that a fixed value of β implies that the whole crowd equally shares the same emotional case. On the other hand, a deeper interpretation of the physics of the crowd suggests to replace this parameter by a variable suitable to describe the dynamics in time and space of emotional states.

In addition, stress is not the only emotional variable which can be taken into account. Different ones can be considered as shown in the next chapters. The presentation in this section is proposed under the assumption that β is a constant parameter, while the challenging problem of modeling the dynamics of emotional states in time and space, which requires a detailed study, is postponed to the next chapters.

The flow patterns can show complex features which have to be taken into account in the modeling approach. For instance, co-existence of dense and rarefied zones have been already shown in Chapter 1. An additional feature is that aggregations of small groups appear and interact each other which corresponds to an endogenous network:

Interactions with the external environment is not simply a matter of quality of a venue, but it might include interactions with various types of vehicles as shown in Fig. 2.1 which corresponds to a complex exogenous network:

These particular features and flow conditions indicate that modeling crowd dynamics cannot be naively developed by simple technical modifications of models of hydrodynamics although specialized at the most appropriate representation and modeling scale. Indeed, it is not surprising that different modeling techniques have been recently proposed and that a commonly shared approach has not yet been achieved.

Bearing all the above in mind, namely without hiding the difficulties we have described above, we tackle the sequential steps of the modeling approach looking ahead to future modeling strategies which will be treated in the last chapter of this Lecture Note.

As mentioned, the first step is the selection of the modeling scale. Accordingly, let us consider the motion of a number N of walkers and define the scales and related variables to be selected to represent the overall state of the system under consideration.

- **Microscopic scale (finite number of degrees of freedom):** The *microscopic description* is given, for each ith walker with $i \in \{1, \ldots, N\}$, by position $\mathbf{x}_i = \mathbf{x}_i(t) = (x_i(t), y_i(t))$ and velocity $\mathbf{v}_i = \mathbf{v}_i(t) = (v_{x,i}(t), v_{y,i}(t))$, where the components $v_{x,i}(t)$ and $v_{y,i}(t)$, (respectively, $x_i(t)$ and $y_i(t)$), of the velocity (respectively, position) are referred to v_M (respectively, to ℓ).

  The independent variable is the dimensionless time t, while the dynamics in space is defined by the positions occupied by all walkers, where their number $N = N(t)$ of walkers can depend on time due to the inlet-outlet dynamics.

- **Mesoscopic scale (kinetic models):** The dependent variable is a probability distribution function $f = f(t, \mathbf{x}, \mathbf{v})$ at time t over the *micro-state* position $\mathbf{x}$ and velocity $\mathbf{v}$. f can be divided by ρ_M, so that it has a physical meaning if it is positive defined and the integral of f over $\mathbf{v}$ takes value in $[0, 1]$. The one-particle representation is used so that f is linked to the so-called test particle (walker) assumed to be representative of the whole system. Time and space are the independent variables, while the state of each walker is the microscopic variable. Macroscopic observable variables are obtained by velocity-weighted moments of the probability distribution.

- **Macroscopic scale (hydrodynamical models):** The dependent variables are the local density $\rho = \rho(t, \mathbf{x})$ and the local mean velocity $\boldsymbol{\xi} = \boldsymbol{\xi}(t, \mathbf{x})$, where ρ is divided by ρ_M and the mean speed ξ is divided by v_M so that in a well-defined venue ρ and ξ have a physical meaning only in $[0, 1]$. The dimensionless time t and space $\mathbf{x}$ are the independent variables as in the case of kinetic models.

This introduction allows us to define precisely the visibility domain which plays an important role in the modeling of interactions and hence in the derivation of equations.

- **Visibility domain:** The visibility domain has analogous physical and geometrical meanings at each scale as it is an arc of circle symmetric with respect to the velocity direction. At the *microscopic scale*, it is denoted by $\Omega_i = \Omega_i(\mathbf{x}_i)$, where it refers to each individual walker i in $\mathbf{x}_i$ with velocity $\mathbf{v}_i$; at the *mesoscopic scale*, it is denoted by $\Omega = \Omega(\mathbf{x})$ referring to the walker, called test walker, representative of the whole system, in $\mathbf{x}$ with velocity $\mathbf{v}$; at the *macroscopic scale*, it is denoted by $\Omega = \Omega(\mathbf{x})$, where it refers to the elementary physical domain in $\mathbf{x}$ with locally averaged velocity $\boldsymbol{\xi}$. The visibility domain can be geometrically reduced by the presence of walls and obstacles which reduce the geometry of the arc of circle as well as by visibility conditions which reduce the effective radius of the arc of circle.

Let us now consider the mathematical structures which have different analytic features at each scale.

- **Micro-scale:** The description of the dynamics at the micro-scale is delivered by a system of ordinary differential equations corresponding to a Newtonian type dynamics as follows:

$$\begin{cases} \dfrac{d\mathbf{x}_i}{dt} = \mathbf{v}_i\,, \\[2mm] \dfrac{d\mathbf{v}_i}{dt} = \mathbf{F}_i(\mathbf{x}_1,\ldots,\mathbf{x}_N,\mathbf{v}_1,\ldots,\mathbf{v}_N\,;\alpha,\beta,\Sigma), \end{cases} \tag{2.1}$$

where $\mathbf{F}_i(\cdot)$ is a psycho-mechanical acceleration acting on the ith walker based on the action of other walkers in his/her visibility/sensitivity zone. Namely, the said acceleration depends on Σ, namely on the overall geometry of the venue, as interactions take into account this feature, which is not accounted for by classical particles until these materially collide with the wall. On the other hand, walkers develop a walking strategy which attempts to avoid their physical contact with the walls.

Hints: The key problem consists in modeling the acceleration terms accounting for the quality and geometry of the venue as well as of the overall crowd distribution in space. Hence, the approach to individual based models is based on the modeling of interactions between each individual with the others. In some cases, only binary interactions are supposed, while collective interactions are not taken into account, while interactions with walls and trend toward the exit are modeled by simple rules.

A model which has been well received by the scientific community is the so-called *social force model* proposed by Helbing and coworkers [17], which has been later developed to model evacuation dynamics [16] and panic situations [15]; see also [14]. In principle, this approach can be generalized to the case of groups moving according to different behaviors, as an example, by including the presence of leaders, as well as multiple interactions. Accounting for the heterogeneity of walkers, which is an important feature of crowd dynamics, still needs additional work to be exhaustively treated.

- **Meso-scale:** The mathematical structures underlying the derivation of models at the mesoscopic scale inspired by the kinetic theory approach models are stated in terms of an evolution equation for the distribution function f, deduced as a balance law in the space of the microscopic states. A basic structure is:

$$\frac{\partial f}{\partial t} + \mathbf{v} \cdot \nabla_{\mathbf{x}} f = G(f, f\,;\alpha,\beta,\Sigma) - f\, L(f\,;\alpha,\beta,\Sigma), \tag{2.2}$$

where $\nabla_{\mathbf{x}}$ denotes the gradient operator with respect to the space variables. In addition, G and L are operators acting on the distribution function f. These operators describe the *gain* and the *loss* of pedestrians in the elementary volume of the phase space about the test microscopic state $(\mathbf{x}, \mathbf{v})$. Their modeling can be achieved by different ways of modeling pedestrian interactions at the microscopic scale.

In more detail, the derivation can be obtained by distinguishing the interacting entities, viewed as active particles, in short a-particles, into three types of them, namely the *test, field*, and the *candidate* a-particles. Their distribution functions are, respectively, $f(t, \mathbf{x}, \mathbf{v})$, $f(t, \mathbf{x}^*, \mathbf{v}^*)$, and $f(t, \mathbf{x}, \mathbf{v}_*)$. The test particle is representative of the whole system, while the candidate particle can acquire, in probability, the micro-state of the test particle after interaction with the field particles. The test particle loses its state by interaction with the field a-particles. A quite general structure refers to papers [3, 6]

$$\left(\frac{\partial}{\partial t} + \mathbf{v} \cdot \nabla_{\mathbf{x}} \right) f(t, \mathbf{x}, \mathbf{v}) = J[f](t, \mathbf{x}, \mathbf{v}; \alpha, \beta, \Sigma)$$

$$= \int_{D_v \times D_v} \int_{\Omega(\mathbf{x})} \eta[f](\mathbf{x}, \mathbf{x}^*, \mathbf{v}_*, \mathbf{v}^*; \alpha, \beta, \Sigma) \, \mathcal{A}[f](\mathbf{v}_* \to \mathbf{v} | \mathbf{x}, \mathbf{x}^*, \mathbf{v}_*, \mathbf{v}^*; \alpha, \beta, \Sigma)$$
$$\times f(t, \mathbf{x}, \mathbf{v}_*) f(t, \mathbf{x}^*, \mathbf{v}^*) \, d\mathbf{x}^* \, d\mathbf{v}_* \, d\mathbf{v}^*$$

$$- f(t, \mathbf{x}, \mathbf{v}) \int_{D_v} \int_{\Omega(\mathbf{x})} \eta[f](\mathbf{x}, \mathbf{x}^*, \mathbf{v}, \mathbf{v}^*; \alpha, \beta, \Sigma) \, f(t, \mathbf{x}^*, \mathbf{v}^*) \, d\mathbf{x}^* \, d\mathbf{v}^*, \qquad (2.3)$$

where D_v defines the velocity domain and $\Omega(\mathbf{x})$ has been defined above.

The following quantities describe the dynamics of interactions.

- The *interaction rate* $\eta[f](\mathbf{x}, \mathbf{x}^*, \mathbf{v}, \mathbf{v}^*; \alpha, \beta, \Sigma)$ (respectively, $\eta[f]$ ($\mathbf{x}, \mathbf{x}^*, \mathbf{v}_*, \mathbf{v}^*; \alpha, \beta, \Sigma$) which models the frequency by which a test (or candidate) particle in $\mathbf{x}$ develops contacts, in the visibility domain, with a field particle.

- The *transition probability density* $\mathcal{A}[f](\mathbf{v}_* \to \mathbf{v} | \mathbf{x}, \mathbf{x}^*, \mathbf{v}_*, \mathbf{v}^*; \alpha, \beta, \Sigma)$ which models the probability density that a candidate particle in $\mathbf{x}$ modifies the velocity into that of the test particle due to the interaction with field particles in the visibility domain $\Omega(\mathbf{x})$ which depends on localization of the particles.

- The *parameters α and β* model, respectively, the quality of the venue and the level of stress, while the interaction terms η and $\mathcal{A}$ depend on these parameters as well as on the overall shape of the venue.

  Hint: The key problem consists in modeling the physical quantities η, $\mathcal{A}$, and Ω. Specific examples, selected among several ones, have been proposed in [3] for models with discrete velocities and in [6] for models with continuous velocity distributions.

The modeling approach has been only recently developed arguably due to the conceptual difficulties to tackle this approach. However, the methodological approach to model living, hence complex, systems proposed in the book [2] has contributed to this challenging aim. Further conceptual contributions have been given by the mathematical theory of evolutionary games [19, 25], while stochastic interactions have been introduced in the book cited above.

- **Macro-scale:** The description of the dynamics at the macro-scale is delivered by a system of partial differential equations corresponding to conservation of mass and linear momentum equilibrium as follows:

$$
\begin{cases}
\partial_t \rho + \nabla_x \cdot (\rho\, \boldsymbol{\xi}) = 0, \\
\\
\partial_t\, \boldsymbol{\xi} + (\boldsymbol{\xi} \cdot \nabla_x)\, \boldsymbol{\xi} = \mathcal{F}[\rho, \boldsymbol{\xi}; \alpha, \beta, \Sigma],
\end{cases}
\tag{2.4}
$$

where $\mathcal{F}[\rho, \boldsymbol{\xi}; \alpha, \beta, \Sigma]$ is a psycho-mechanical acceleration acting on walkers.

Hints: The key problem consists in modeling the acceleration terms accounting for the quality and geometry of the venue as well as the overall crowd distribution in space. The modeling at the macroscopic scale, which is developed by methods analogous to those of hydrodynamics, is reviewed in the survey [18]. The challenging conceptual difficulty consists in understanding how the crowd, viewed as a continuum, selects the velocity direction and the speed by which pedestrians move. The approach proposed in [18] suggests to compute optimal trajectories by which pedestrians reach the exit. Detailed calculations are made in steady flow conditions. More complexity appears to be the problem of modeling irrational behaviors in panic situations. Additional results are reviewed in the book [11] which provides an exhaustive survey of the existing literature in the field mainly focused at the microscopic and macroscopic scale.

2.3   MODELING STRATEGY AND VALIDATION

As shown in the preceding section, different mathematical structures correspond to each one of the three modeling scales therein reported. Nevertheless, a general modeling strategy can be designed, where the first step is the selection of the most appropriate scale to be chosen with the aim of accounting for the complexity features of human crowds introduced in Section 1.3. An additional requirement, which can also contribute to the selection of the most appropriate scale, is the derivation of models suitable to reproduce empirical data without any artificial insertion of them into models.

Bearing this introductory note in mind, let us first propose the modeling strategy and subsequently introduce some general ideas on the problem of validation of models.

1. Assessment of the complexity features of human crowds, namely ability to express a strategy and self-organization; heterogeneity at the micro-scale, namely different walking abilities; capacity to learn from past experience; nonlinearly additive and nonlocal interactions; and sensitivity to quality and geometry of the venues where the crowd moves.

2. Heterogeneity at the macro-scale, for instance subdivision into groups with different walking objectives or the possible presence of leaders.

3. Selection of the social phenomena to be inserted in the model for instance stress conditions which are highly influenced by perception of crisis situations or even presence of antagonist groups.

4. Assessment of the geometrical and quality features of the venue where the crowd moves, where the former refers to narrow corridors, large rooms, presence of up-going and down-going steps, etc., while the latter refers to physical features such as lightening, properties of the floor, presence of smoke, and visibility.

5. Selection of the modeling scale and derivation of a mathematical structure are consistent with the requirements in the first four items; due to the aforementioned subdivisions, the structure should correspond to a mixture of different groups.

6. Derivation of models by inserting, into the said structure, the mathematical description of interactions for both social and mechanical dynamics including their reciprocal interplay.

7. Tuning of the model by identification of the parameters which characterize the model and validation of models by appropriate comparisons with empirical data.

In principle, this modeling strategy can be developed at each scale also in view of the selection of the most appropriate as we shall see in the next section. Therefore, let us preliminarily examine the key problem of model validation, while this topic will be made more precise in Section 2.4 referring specifically to well-defined models. A possible validation strategy has been elaborated in [5], basically related to the following three main abilities required by crowd models:

1. capturing the complexity features of human crowds viewed as a living, hence complex, system;

2. ability to reproduce "quantitatively" of the velocity and fundamental diagrams of crowd flows, namely mean velocity and flow vs. local density;

3. ability to reproduce "qualitatively" emerging behaviors which are observed in experiments, such as the creation of lanes in narrow streets and increasing high walker concentration, and even evacuation time, in stressful conditions; and

4. account for environmental conditions which can determine different observable dynamics, e.g., different velocity and fundamental diagrams, as well as emerging behaviors.

These four items show that we need information from experiments. However, the literature on empirical data on crowd dynamics still needs to be exhaustively developed although it is rapidly increasing due to the growing interest of the scientific community and of the society on this topic. Indeed, the amount of available empirical data is not yet sufficient to lead to a fully satisfactory validation process. In more detail, the following types of empirical data can be recovered.

- *Velocity diagrams:* These diagrams [33, 34] depend also on the quality of the area where the crowd moves. A possible way to address this issue consists of introducing a parameter related to the quality of the venue where walkers move. High values of this parameter correspond to high quality which allows high speed, while low values correspond to low quality which can even prevent motion. Additional bibliography enlightened by sharp reasonings on the use of empirical data toward design and tuning of models is reported in [32].

- *Walking behaviors:* Empirical results have been proposed in [12, 23, 24], which can contribute to the derivation of models at the microscopic scale as well as to the dynamics of interactions to derive models at the mesoscopic scale.

- *Modeling the role of environmental conditions on walking behaviors:* For instance, the presence of smoke, fire, and environmental conditions to be accounted for by appropriate parameters [30].

- *Statistical tools* toward the tuning of parameters of the model [9, 10] by minimizing the distance between prediction of models and observed empirical data.

- *Technological devices* toward the individual identification of walkers [22, 28] leading to empirical data useful for validation.

In addition, some recent contributions have proposed specific techniques toward the tuning of models, namely parameter assessment, in [9, 10], where one of the main difficulties to be tackled is that the greatest part of empirical data sets are available at the macroscopic scale, while the modeling process needs a detailed understanding of the dynamics at the microscopic scale.

Focusing on the fundamental diagrams, it is worth stressing that their artificial insertion into the equations of the model takes computational models far from the real physics of the system. Indeed, velocity diagrams must be reproduced, as a consequence of interactions at the microscopic scale, and not implemented in the model.

2.4   ON THE SELECTION OF THE MODELING SCALE

Let us now focus on the selection of the most appropriate modeling scale. Indeed, mesoscopic scale appears to be as the most appropriate to capture the complexity features of human crowd which have been presented in Chapter 1. These features play a key role in the study of crowd dynamics in extreme, high-risk situations which enhance them.

Based also on our professional experience, which has been enriched to the large European network:

> European Unions – Seventh Framework Program, **eVACUATE: A holistic, scenario independent, situation awareness and guidance system for sustaining the Active Evacuation Route for large crowds**, May 2013 – July 2017,

we support the choice of the kinetic theory approach as the most appropriate to account for the complexity features of human crowds. Some work has been done on this topic [7], so that we can propose the following items to enforce this choice.

- The heterogeneity of walkers can be taken into account by the representation of the system by a probability distribution over the micro-state of each individual. The heterogeneity is enhanced in crisis situations which can amplify the gap between individuals keen to strong emotional states and individuals who are able to keep rational behaviors.

- The use of theoretical tools of game theory in the modeling of interactions allows for the inclusion of all possible trends of walkers interacting with each other and with the venue where they walk. As we shall see, walkers are subject to different trends, for instance search for the exit direction, desire to avoid overcrowded areas and walls or obstacles, but also attraction to the main stream. Theoretical tools of game theory provide a weighed selection of all these tools. The weight of this selection depends on the levels of emotional states which induce the onset of irrational behaviors out of otherwise rational ones.

- Subdivision into functional subsystems allows for the inclusion of different typologies of walkers. For instance, individuals who, in crisis situations, need physical support to evacuate or leaders who can exhibit rational behaviors and attract the whole crowd to safe exit routes.

- Kinetic models have shown in [7] the ability to reproduce, in steady uniform flow conditions, the so-called velocity and fundamental diagrams, namely mean speed and flow vs. density. This achievement has been obtained as an emerging behavior induced by interactions and self-organization without imposing it directly into the model.

- As mentioned, mathematical models are required to reproduce emerging behaviors which are observed under some specific initial and boundary conditions. A specific example, studied in [6], is the self-organization in two counter flows moving in long corridors even if they are initially randomly distributed. The application has shown that the crowd segregates into two main streams: one on the left moving along one of the directions and one stream in the opposite direction. In addition, two numerically minor streams move along the walls in directions opposite to that of the two main streams.

- Recent literature has been devoted to modeling contagion in moving crowds [21]. This approach can be related to emerging patterns of infection [4].

A more detailed description of the above items will be given in Chapter 4, referring to well-defined models. However, this selection does not hide that each scale has some advantages and drawbacks. For instance, modeling at the microscopic scale is consistent with the specific feature of crowds which are systems with finite number of degrees of freedom, but it is technically difficult accounting for the heterogeneity of individuals and dealing with a huge number of equations which appear in huge crowds with thousands of individuals.

The modeling at the macro-scale shows the advantage of providing directly the macroscopic observable quantities which are needed to study the flow of crowds by equations which can be treated numerically by finite volumes even in complex venues. However, heterogeneity is lost in the local averaging process necessary to derive the said equations. An additional problem is that the continuity assumption, which is typical of the hydrodynamic approach, is not consistent with physical reality of a multi-particle system where rarefied flows coexist with almost continuous flows. This topic is one of the research perspectives discussed in Chapter 5.

These reasonings motivate possible developments of a multiscale vision which has been proposed in [1], where models at each scale are derived by the same rationale and they include parameters related to the same specific features selected in the modeling approach.

2.5   TOWARD MODELING PERSPECTIVES

The contents of this chapter have been addressed to multiscale topics first by defining the different mathematical structures underlying the derivation of models and subsequently by selecting the most appropriate modeling scale, where the selection has been made by investigating how far the kinetic theory approach can capture the complexity features of human crowds. This selection has been proposed without forgetting that a multiscale vision of crowd modeling is an important target to pursue and that all possible scales should technically interact.

Therefore, this chapter opens to the contents of Chapter 3, where mathematical structures, more specialized than (2.3), can be designed accounting for the various possible application of models. These structures can be related to those of the classical kinetic theory toward of a deep understanding of the substantial difference between the kinetic theory of active particles and the theory of classical particles.

This chapter has also proposed some guidelines for the validation of models which will be applied in Chapter 4 to validate specific models. This delicate issue, which precedes the application of models, has to face the additional difficulty that a unified analysis of emerging behaviors. A valuable presentation of a variety of emerging behaviors has been delivered in the excellent book by Kerner [20]. However, an analogous contribution in the case of crowd dynamics still appears to be a challenging research perspective, but the number of papers which provide empirical data that can effectively help in the modeling approach is rapidly growing, for instance [9, 10, 23, 24, 28] and various others which are not reported here.

As mentioned in Section 2.2, we do not naively claim that the models presented in this chapter provide an exhaustive approach to the target to cast into a mathematical framework a complex system which requires additional work and perspective ideas. Nevertheless, we consider our presentation consistent with an advanced stage of the research activity in the field. The last chapter is devoted, as we shall see, to investigate the aforementioned perspectives.

2.6   BIBLIOGRAPHY

[1]   B. Aylaj, N. Bellomo, L. Gibelli, and A. Reali, On a unified multiscale vision of behavioral crowds, *Mathematical Models and Methods in Applied Sciences*, 30:1–22, 2020. DOI: 10.1142/S0218202520500013 18, 28

[2]   N. Bellomo, A. Bellouquid, L. Gibelli, and N. Outada, *A Quest Towards a Mathematical Theory of Living Systems*, Birkhäuser-Springer, New York, 2017. DOI: 10.1007/978-3-319-57436-3 24

[3]   N. Bellomo, A. Bellouquid, and D. Knopoff, From the micro-scale to collective crowd dynamics, *Multiscale Modelling and Simulation*, 11:943–963, 2013. DOI: 10.1137/130904569 23

[4]   N. Bellomo, R. Bingham, M. A. J. Chaplain, G. Dosi, G. Forni, D. A. Knopoff, J. Lowengrub, R. Twarock, and M. E. Virgillito, A multi-scale model of virus pandemic: Heterogeneous interactive entities in a globally connected world, *Mathematical Models and Methods in Applied Sciences*, 20, 2020. DOI: 10.1142/s0218202520500323 28

[5]   N. Bellomo, D. Clarke, L. Gibelli, P. Townsend, and B. J. Vreugdenhil, Human behaviours in evacuation crowd dynamics: From modelling to "big data" toward crisis management, *Physics of Life Reviews*, 18:1–21, 2016. DOI: 10.1016/j.plrev.2016.05.014 25

[6]   N. Bellomo and L. Gibelli, Toward a behavioral-social dynamics of pedestrian crowds, *Mathematical Models and Methods in Applied Sciences*, 25:2417–2437, 2015. DOI: 10.1142/S0218202515400138 23, 24, 27

[7]   N. Bellomo and L. Gibelli, Behavioral crowds: Modeling and Monte Carlo simulations toward validation, *Computers and Fluids*, 141:13–21, 2016. DOI: 10.1016/j.compfluid.2016.04.022 19, 27

[8]   N. Bellomo, L. Gibelli, and N. Outada, On the interplay between behavioral dynamics and social interactions in human crowds, *Kinetic and Related Models*, 12:397–409, 2019. DOI: 10.3934/krm.2019017 18

[9] A. Corbetta, L. Bruno, A. Mountean, and F. Yoschi, High statistics measurements of pedestrian dynamics, models via probabilistic method, *Transportation Research Proceedings*, 2:96–104, 2014. DOI: 10.1016/j.trpro.2014.09.013 26, 29

[10] A. Corbetta, A. Mountean, and K. Vafayi, Parameter estimation of social forces in pedestrian dynamics models via probabilistic method, *Mathematical Biosciences Engineering*, 12:337–356, 2015. DOI: 10.3934/mbe.2015.12.337 26, 29

[11] E. Cristiani, B. Piccoli, and A. Tosin, *Multiscale Modeling of Pedestrian Dynamics*, Springer Italy, 2014. DOI: 10.1007/978-3-319-06620-2 24

[12] W. Daamen and S. P. Hoogedorn, Experimental research of pedestrian walking behavior, *TRB Annual Meeting CD-ROM*, 2006. DOI: 10.3141/1828-03 26

[13] L. Gibelli and N. Bellomo, Eds., *Crowd Dynamics Volume 1: Theory, Models, and Safety Problems*, Birkhäuser, New York, 2018. DOI: 10.1007/978-3-030-05129-7

[14] D. Helbing, Traffic and related self-driven many-particle systems, *Review Modern Physics*, 73:1067–1141, 2001. DOI: 10.1103/revmodphys.73.1067 22

[15] D. Helbing, I. Farkas, and T. Vicsek, Simulating dynamical feature of escape panic. *Nature*, 407:487–490, 2000. DOI: 10.1038/35035023 22

[16] D. Helbing and A. Johansson, Pedestrian crowd and evacuation dynamics, *Encyclopedia of Complexity and System Science*, pages 6476–6495, Springer, 2009. DOI: 10.1007/978-3-642-27737-5_382-5 22

[17] D. Helbing and P. Molnár, Social force model for pedestrian dynamics, *Physics Review E*, 51:4282–4286, 1995. DOI: 10.1103/physreve.51.4282 22

[18] R. L. Hughes, The flow of human crowds, *Annual Review Fluid Mechanics*, 35:169–182, 2003. DOI: 10.1016/s0378-4754(00)00228-7 24

[19] J. Hofbauer and K. Sigmund, Evolutionary game dynamics, *Bulletin American Mathematical Society*, 40:479–519, 2003. DOI: 10.1090/s0273-0979-03-00988-1 24

[20] B. Kerner, *The Physics of Traffic*,   Springer, New York, Heidelberg, 2004. DOI: 10.1007/978-3-540-40986-1 29

[21] D. Kim and A. Quaini, A kinetic theory approach to model pedestrian dynamics in bounded domains with obstacles, *Kinetic and Related Models*, 12:1273–1296, 2019. DOI: 10.3934/krm.2019049 28

[22] A. Kirchner and A. Schadschneider, Simulation of evacuation processes using a bionics-inspired cellular automaton model for pedestrian dynamics, *Physics A*, 312:260–276, 2002. DOI: 10.1016/s0378-4371(02)00857-9 26

[23] M. Moussaid, D. Helbing, S. Garnier, A. Johanson, M. Combe, and G. Theraulaz, Experimental study of the behavioral underlying mechanism underlying self-organization in human crowd, *Proc. of the Royal Society B: Biological Sciences*, 276:2755–2762, 2009. DOI: 10.1098/rspb.2009.0405 26, 29

[24] M. Moussaïd and G. Theraulaz, Comment les piétons marchent dans la foule, *La Recherche*, 450:56–59, 2011. 26, 29

[25] M. A. Nowak, *Evolutionary Dynamics. Exploring the Equations of Life*, Harvard University Press, 2006. DOI: 10.2307/j.ctvjghw98 24

[26] X. Li, F Guo, H Kuang, Z Geng, and Y. Fan, An extended cost potential field cellular automaton model for pedestrian evacuation considering the restriction of visual field, *Physica A*, 515:47–56, 2019. DOI: 10.1016/j.physa.2018.09.145 18

[27] X. Li, F. Guo, H. Kuang, and H. Zhou, Effect of psychological tension on pedestrian counter flow via an extended cost potential field cellular automaton model, *Physica A*, 487:47–56, 2017. DOI: 10.1016/j.physa.2017.05.070 18

[28] D. Roggen, M. Wirz, G. Tröster, and D. Helbing, Recognition of crowd behavior from mobile sensors with pattern analysis and graph clustering methods, *Networks Heterogenius Media*, 6:521–544, 2011. DOI: 10.3934/nhm.2011.6.521 26, 29

[29] E. Ronchi, Disaster management: Design buildings for rapid evacuation, *Nature*, 528(7582):333, 2015. DOI: 10.1038/528333b.

[30] E. Ronchi, S. M. V. Gwynne, D. A. Purser, and P. Colonna, Representation of the impact of smoke on agent walking speeds in evacuation models, *Safety Science*, 52:28–36, 2013. DOI: 10.1007/s10694-012-0280-y 26

[31] E. Ronchi, E. D. Kuligowski, D. Nilsson, R. D. Peacock, and P. A. Reneke, Assessing the verification and validation of building fire evacuation models, *Fire Technology*, 52(1):197–219, 2016. DOI: 10.1007/s10694-014-0432-3

[32] A. Schadschneider, M. Chraibi, A. Seyfried, A. Tordeux, and J. Zhang, Pedestrian dynamics: From empirical results to modeling, in *Crowd Dynamics Volume 1: Theory, Models, and Safety Problems*, L. Gibelli and N. Bellomo, Eds., pages 63–102, Birkhäuser, New York, 2018. DOI: 10.1007/978-3-030-05129-7_4 26

[33] A. Schadschneider and A. Seyfried, Empirical results for pedestrian dynamics and their implications for cellular automata models, in *Pedestrian Behavior-Models, Data Collection and Applications*, H. Timmermans, Ed., Chapter 2, pages 27–44, Emerald Group Publishing, 2009. DOI: 10.1108/9781848557512-002 26

[34]   A. Schadschneider and A. Seyfried,   Empirical results for pedestrian dynamics and their implications for modeling, *Networks Heterogenous Media* 6:545–560, 2011. DOI: 10.3934/nhm.2011.6.545 26

CHAPTER 3

From Classical Kinetic Theory to Active Particle Models

Abstract   This chapter presents the mathematical tools which can be used toward the modeling the dynamics of human crowds. The presentation is in three parts. First, a concise description of the classical Boltzmann equation and of the main properties of the model is proposed. Next, some models derived from specific simplifications or generalizations of the original model are presented. Finally, the mathematical tools and properties of the kinetic theory of active particles are proposed in view of the modeling applications treated in the following chapters. The contents are essentially theoretical, but special focus is posed to properties of the mathematical structures which can used in crowds modeling.

3.1   INTRODUCTION

This chapter is devoted to the derivation of mathematical tools suitable to define the general mathematical framework which can support the kinetic theory approach to modeling crowd dynamics. Although the physics of human crowds is far away from the theory of classical particles, models of the kinetic theory of diluted gases and, in particular, the celebrated Boltzmann equation, are essential references. Hence, this chapter starts from a concise presentation of these models.

In addition to the classical Boltzmann equation, some particular models, however related to the general equation, can inspire the approach to crowd modeling. In detail, this chapter provides a presentation of models of this type from the so-called BGK models, where the Boltzmann's so-called collision operator is replaced by a phenomenological model describing a linear or nonlinear trend to equilibrium to the discrete Boltzmann equation, where particles are allowed to attain only a finite number of velocities. Some features of these models have already shown their ability to contribute to the derivation of models of crowd dynamics, as shown in [3, 11].

The presentation of these models does not hide that the mathematical structures underlying the derivation of kinetic models of human crowds present a substantial difference with respect to the mathematical kinetic theory of classical particles. A detailed study of these differences can contribute to enlighten us as to how different models of crowd dynamics can be derived.

This chapter aims at showing the bridge linking the two related, however different, kinetic theories. The contents is as follows.

Section 3.2 provides a concise presentation of the classical Boltzmann equation focusing on the phenomenological derivation and on the main properties of this celebrated model of mathematical physics. As mentioned, only a concise introduction is proposed, while the interested reader can refer to Chapter 2 of [2] for a more extended overview and to [7] for an exhaustive overview. Section 3.3 describes some models of the classical kinetic theory derived, as mentioned above, under technical assumptions that simplify the physical reality of classical particles. For instance, the so-called discrete Boltzmann equations, the so-called BGK models, the Enskog, and the Vlasov-type models where interactions are smeared in the whole space and discrete velocity models. Section 3.4 presents the mathematical structures of the so-called kinetic theory of active particles, where the microscopic state includes not only mechanical variables, typically position and velocity, but also additional variables deemed to model the "social" properties of individuals in the crowd. The main difference, as we shall see, is that interactions, which are nonlinear and nonlocal, do not obey rules of classical mechanics as the social variable modifies these rules. Section 3.5 proposes a critical analysis suitable to discuss again the substantial difference between classical and active particles in view of the derivation of specific models.

3.2   THE BOLTZMANN EQUATION

The Boltzmann equation describes the collective dynamics of a large system composed by spherically symmetrical classical particles modeled as point masses, where their microscopic state is given by position $\mathbf{x} \in \mathbb{R}^3$ and velocity $\mathbf{v} \in \mathbb{R}^3$. The derivation is developed under the hypothesis that the mean-free path is large enough that only binary collisions are important, namely the probability of multiple collisions is not significant. Boltzmann's great idea has been the representation of the system by a probability distribution over the microscopic state rather than by position and velocity of all particles. The celebrated equation leads to the description of the said probability distribution, while macroscopic observable quantities are obtained as velocity weighted moments of the said distribution function.

The framework needs a stochastic representation of an N-particle system by the probability distribution over the microscopic state of all N particles

$$f_N = f_N(t, \mathbf{x}_1, \ldots, \mathbf{x}_N, \mathbf{v}_1, \ldots, \mathbf{v}_N),$$

such that $f_N \, d\mathbf{x}_1 \ldots d\mathbf{x}_N d\mathbf{v}_1 \ldots d\mathbf{v}_N$ represents the probability of finding, at time t, a particle in the elementary volume of the N-particles phase space, while the microscopic state at time t is the vector variable defined by all positions and velocities.

If the number of particles is constant in time and if the dynamics follow Newton's rules, a continuity equation in the phase space is satisfied. Formal calculations, related to conservation of particles, yield

$$\partial_t f_N + \sum_{i=1}^{N} \mathbf{v}_i \cdot \nabla_{\mathbf{x}} f_N + \frac{1}{m} \sum_{i=1}^{N} \nabla_{\mathbf{x}}(\mathbf{F}_i \cdot f_N) = 0, \tag{3.1}$$

where $\mathbf{F}_i = \mathbf{F}_i(t, \mathbf{x})$ is the force applied to each particle by the overall system. This equation, which is known as "Liouville equation," is not practical as it involves a too large number of variables.

This difficulty has been tackled by Boltzmann's theory which is presented in the next three subsections focused on:

(i) representation of an N-particle system;

(ii) derivation of the equation; and

(iii) analysis of the main properties of the equation.

The presentation, as mentioned, is concise having been addressed to the appropriate literature. Some key remarks are inserted within the presentation in view of a critical analysis on the differences between the classical kinetic theory and the theory of active particles.

The application of the Boltzmann equation to the study of fluid dynamical problems has a long story documented in various books, for instance [1]. These applications have generated a variety of challenging analytic and computational problems which are not treated in this Lecture Note which are mainly devoted to the kinetic theory of active particles.

3.2.1 STATISTICAL REPRESENTATION

The idea by Ludwig Boltzmann has been the description of the overall state by a continuous probability distribution over the microscopic state of the particles, namely position and velocity of the *test particle* representative of the whole. Therefore, he introduced the one particle distribution function

$$f = f(t, \mathbf{x}, \mathbf{v}) : \quad [0, T] \times \mathbb{R}^3 \times \mathbb{R}^3 \rightarrow \mathbb{R}_+,$$

such that under suitable integrability conditions $f(t, \mathbf{x}, \mathbf{v})d\mathbf{x}\,d\mathbf{v}$ defines the number of particles that at time t are in the elementary volume $[\mathbf{x}, \mathbf{x} + d\mathbf{x}] \times [\mathbf{v}, \mathbf{v} + d\mathbf{v}]$, called phase space, of the microscopic states.

Suppose now that $\mathbf{v}^r f(t, \mathbf{x}, \mathbf{v}) \in L_1(\mathbb{R}^3)$, for $r = 0, 1, 2, \ldots$, then macro-scale quantities are obtained by weighted moments as follows:

$$M_r = M_r[f](t, \mathbf{x}) = \int_{\mathbb{R}^3} m\, \mathbf{v}^r\, f(t, \mathbf{x}, \mathbf{v})\, d\mathbf{v}, \tag{3.2}$$

where m is the mass of the particles, $r = 0$ corresponds to the local density $\rho(t, \mathbf{x})$, and $r = 1$ to the linear momentum $\mathbf{Q}(t, \mathbf{x})$, while $r = 2$ to the mechanical energy. Therefore, the local density and mean velocity are computed as follows:

$$\rho = \rho(t, \mathbf{x}) = m \int_{\mathbb{R}^3} f(t, \mathbf{x}, \mathbf{v}) \, d\mathbf{v}, \tag{3.3}$$

$$\boldsymbol{\xi} = \boldsymbol{\xi}(t, \mathbf{x}) = \frac{1}{\rho(t, \mathbf{x})} \int_{\mathbb{R}^3} m \, \mathbf{v} \, f(t, \mathbf{x}, \mathbf{v}) \, d\mathbf{v}, \tag{3.4}$$

while the kinetic translational energy is given by the second-order moment

$$\mathcal{E} = \mathcal{E}(t, \mathbf{x}) = \frac{1}{2\rho(t, \mathbf{x})} \int_{\mathbb{R}^3} m \, |\mathbf{v}|^2 \, f(t, \mathbf{x}, \mathbf{v}) \, d\mathbf{v}, \tag{3.5}$$

where $|\mathbf{v}|$ is the velocity modulus or speed. The energy expression (3.5) can be related, equilibrium conditions to the local temperature based on the principle of repartition of energy, which is valid only in the said conditions.

3.2.2   ON THE PHENOMENOLOGICAL DERIVATION OF THE BOLTZMANN EQUATION

The derivation of the equation, which is occasionally called *formal* as it requires ad hoc phenomenological assumptions, can be developed through the following sequential steps:

1. development of detailed calculations of the interaction dynamics of perfectly elastic pair of spherically shaped particles;

2. computation of the net flow of particles entering into the elementary volume $[\mathbf{x} + d\mathbf{x}] \times [\mathbf{v} + d\mathbf{v}]$ of the phase space and leaving it due to the interaction (collision) dynamics; and

3. modeling the time and space dynamics, in the said elementary volume, for the one particle distribution function by equating the local evolution of the distribution function to the net flow of particles due to interactions.

Let us now consider the *interaction dynamics* for binary encounters preserving mass, linear momentum, and energy between two particles with equal mass. Moreover, let us denote by $\mathbf{v}$ and $\mathbf{v}_*$ the velocities of the two interacting particles before interaction and by $\mathbf{v}'$ and $\mathbf{v}'_*$ their velocities after interaction, respectively. The conservation of the aforementioned mechanical quantities implies

$$\mathbf{v} + \mathbf{v}_* = \mathbf{v}' + \mathbf{v}'_*, \tag{3.6}$$

and

$$|\mathbf{v}|^2 + |\mathbf{v}_*|^2 = |\mathbf{v}'|^2 + |\mathbf{v}'_*|^2. \tag{3.7}$$

This system of four scalar equations is not sufficient to compute the six components of the post-interaction velocities. Usually, see [7]; the post-interaction velocities are related to the pre-interaction ones as follows:

$$\mathbf{v}' = \mathbf{v} + \mathbf{k}(\mathbf{k} \cdot \mathbf{q}), \quad \text{and} \quad \mathbf{v}'_* = \mathbf{v}_* - \mathbf{k}(\mathbf{k} \cdot \mathbf{q}), \tag{3.8}$$

where $\mathbf{q} = \mathbf{v} - \mathbf{v}_*$ is the relative velocity and $\mathbf{k}$ the whole set of interactions which is obtained by integration of $\mathbf{k}$ over the domain

$$\mathcal{K} = \left\{ \mathbf{k} \in \mathbb{R}^2 \ : \ ||\mathbf{k}|| = 1, \mathbf{k} \cdot \mathbf{q} \ \geq 0 \right\},$$

is the unit vector directed along the line joining the centers of the spheres.

Remark 3.1   We have chosen the term *interaction* rather than *collision*, which is often used in the literature, as particles do not really collide as repulsive forces modify their trajectories similarly to a collision, but not precisely as in the case of billiard balls.

The derivation of the equation is obtained, as mentioned, by a balance of particles in the elementary volume of the phase space $[\mathbf{x}, \mathbf{x} + d\mathbf{x}] \times [\mathbf{v} + d\mathbf{v}]$, which can be formally obtained by equating the flow due to transport to the net flow due to interactions, as follows:

$$\begin{aligned}(\partial_t + \mathbf{v} \cdot \nabla_{\mathbf{x}} + \mathbf{F}(t, \mathbf{x}) \cdot \nabla_{\mathbf{v}}) \, f(t, \mathbf{x}, \mathbf{v}) &= \mathcal{N}[f, f](t, \mathbf{x}, \mathbf{v}) \\ &= G(f, f)(t, \mathbf{x}, \mathbf{v}) - L(f, f)(t, \mathbf{x}, \mathbf{v}),\end{aligned} \tag{3.9}$$

where $\mathcal{N}$ is the net flow, while G and L denote the so-called *gain* and *loss* terms amounting to the inlet particles due to interactions into the elementary volume $d\mathbf{x} \, d\mathbf{v}$ of the phase space and, respectively, the outlet from the said volume.

The calculation of these quantities needs some heuristic assumptions.

1. The probability of occurrence of interactions involving more than two particles is negligible with respect the probability corresponding to binary encounters.

2. The effect of external forces on the molecules during the mean interaction time is negligible with respect to the interacting molecular forces.

3. Interactions are not dissipative, namely mass and momentum, are preserved.

4. The asymptotic pre-interaction velocities of two molecules are not correlated as well as their post-interaction velocities. This hypothesis is known as the molecular chaos assumption and implies that the joint particle distribution functions of the two interacting particles can be factorized, namely given by the product $f(t, \mathbf{x}, \mathbf{v}) f(t, \mathbf{x}, \mathbf{v}_*)$.

5. The distribution function does not change very much over a time interval which is larger than the mean interaction time but smaller than the mean free time. Likewise, the distribution function does not change very much over a distance of the order of the range of the intermolecular forces.

Bearing all the above in mind, the terms G and L can be computed explicitly according to the aforementioned assumptions. Technical calculations are reported in specialized books, see [7] and Chapter 2 in [2]. The result is as follows:

$$G(f, f)(t, \mathbf{x}, \mathbf{v}') = d\mathbf{x} \, d\mathbf{v}' \int f(t, \mathbf{x}, \mathbf{v}') f(t, \mathbf{x}, \mathbf{v}'^*) \, \mathbf{q}' \, b' \, db' \, d\varepsilon' \, d\mathbf{v}'^*, \qquad (3.10)$$

and

$$L(f, f)(t, \mathbf{x}, \mathbf{v}) = d\mathbf{x} \, d\mathbf{v} \int f(t, \mathbf{x}, \mathbf{v}) f(t, \mathbf{x}, \mathbf{v}_*) \, \mathbf{q} \, b \, db \, d\varepsilon \, d\mathbf{v}_*, \qquad (3.11)$$

where b and ε define, in cylindrical coordinates, the relative position of the interacting particles and where for simplicity of notations we have not indicated the domains (standard in the classical kinetic theory) of integration.

In these expressions of the collision operators G and L, the following properties have been used:

(i) the modulus of the Jacobian for the equations that relate the post- and pre-collision asymptotic velocities is equal to one, i.e., $d\mathbf{v}' d\mathbf{v}'_* = d\mathbf{v} d\mathbf{v}_*$; and

(ii) due to the energy conservation law, the modulus of the pre- and post-collision relative velocities are equal to each other, i.e., $\mathbf{q}' = \mathbf{q}$.

Therefore, the Boltzmann equation writes as follows:

$$(\partial_t + \mathbf{v} \cdot \nabla_{\mathbf{x}} + \mathbf{F}(t, \mathbf{x}) \cdot \nabla_{\mathbf{v}}) \, f(t, \mathbf{x}, \mathbf{v})$$
$$= \int f(t, \mathbf{x}, \mathbf{v}') f(t, \mathbf{x}, \mathbf{v}'^*) \, \mathbf{q}' \, b' \, db' \, d\varepsilon' \, d\mathbf{v}'^*$$
$$- \int f(t, \mathbf{x}, \mathbf{v}) f(t, \mathbf{x}, \mathbf{v}_*) \, \mathbf{q} \, b \, db \, d\varepsilon \, d\mathbf{v}_*. \qquad (3.12)$$

An alternative to this expression of the Boltzmann equation, which is somehow closer to the mathematical structures of the theory of the active particles is the following:

$$(\partial_t + \mathbf{v} \cdot \nabla_{\mathbf{x}} + \mathbf{F}(t, \mathbf{x}) \cdot \nabla_{\mathbf{v}}) \, f(t, \mathbf{x}, \mathbf{v})$$
$$= \int f(t, \mathbf{x}, \mathbf{v}') f(t, \mathbf{x}, \mathbf{v}'_*) \, W(\mathbf{v}' \to \mathbf{v}'_* | \mathbf{v}, \mathbf{v}_*) \, d\mathbf{v}_* \, d\mathbf{v}'_* \, d\mathbf{v}'$$
$$- \int f(t, \mathbf{x}, \mathbf{v}) f(t, \mathbf{x}, \mathbf{v}_*) \, W(\mathbf{v} \to \mathbf{v}_* | \mathbf{v}', \mathbf{v}'_*) \, d\mathbf{v}_* \, d\mathbf{v}' \, d\mathbf{v}'_*, \qquad (3.13)$$

where W denotes the probability density over the output of the collision being conditioned by the input velocities.

3.2.3   PROPERTIES OF THE BOLTZMANN EQUATION

Although the Boltzmann equation is derived under various approximations of physical reality, it still retains some important features of the physics of system composed of many interacting classical particles. Let us first observe that the local density and mean velocity are computed by

Eqs. (3.3) and (3.4), while the local temperature can be related to the kinetic energy by assuming that the principle of repartition of the energy for a perfect gas can be applied.

Bearing this in mind, some important properties can be mentioned.

1. The collision operator $\mathcal{N}[f, f]$ admits collision invariants corresponding to mass, momentum, and energy, namely:

$$< \phi_r, \mathcal{N}[f, f] >= 0, \quad \text{for} \quad r = 0, 1, 2, \quad \text{with} \quad \phi_r = \mathbf{v}^r, \tag{3.14}$$

where the notation $<, >$ is used for product under the integral sign. Equation (3.14) states conservation of mass, linear momentum, and energy.

2. A unique solution f_e to the equilibrium equation $\mathcal{N}[f, f] = 0$ exists and is the Maxwellian distribution:

$$f_e(\mathbf{x}, \mathbf{v}) = \frac{\rho(t, \mathbf{x})}{[2\pi(k/m)\Theta(\mathbf{x})]^{3/2}} \exp\left\{ -\frac{|\mathbf{v} - \boldsymbol{\xi}(\mathbf{x})|^2}{2(k/m)\Theta(\mathbf{x})} \right\}, \tag{3.15}$$

where k is the Boltzmann constant, and Θ is the temperature given, for a mono-atomic gas, by the approximation of a perfect gas, namely:

$$\mathcal{E} = \frac{3}{2} k\,\Theta,$$

where $\mathcal{E}$ is the already defined kinetic energy.

The first item states conservation of mass

$$\int f(t, \mathbf{x}, \mathbf{v})\, d\mathbf{x}\, d\mathbf{v} = \int f_0(\mathbf{x}, \mathbf{v})\, d\mathbf{x}\, d\mathbf{v},$$

linear momentum

$$\int \mathbf{v}\, f(t, \mathbf{x}, \mathbf{v})\, d\mathbf{x}\, d\mathbf{v} = \int \mathbf{v}\, f_0(\mathbf{x}, \mathbf{v})\, d\mathbf{x}\, d\mathbf{v},$$

and energy

$$\int |\mathbf{v}|^2\, f(t, \mathbf{x}, \mathbf{v})\, d\mathbf{x}\, d\mathbf{v} = \int |\mathbf{v}|^2\, f_0(\mathbf{x}, \mathbf{v})\, d\mathbf{x}\, d\mathbf{v},$$

where $f_0(\mathbf{x}, \mathbf{v}) = f(t = 0, \mathbf{x}, \mathbf{v})$.

Trend to equilibrium is assured by the *kinetic entropy*, namely the H-Boltzmann functional:

$$H[f](t) = \int_{\mathbb{R}^3 \times \mathbb{R}^3} f \, \log(f)(t, \mathbf{x}, \mathbf{v})\, d\mathbf{x}\, d\mathbf{v}, \tag{3.16}$$

which, in the spatially homogeneous case, is monotone decreasing along the solutions and is equal to zero at $f = f_e$. Therefore the Maxwellian distribution with parameters ρ, $\mathbf{U}$, and Θ minimizes H (this is the celebrated H-Theorem [7]).

Mathematical problems for the Boltzmann equation can be classified as the initial value (Cauchy) problem in unbounded domains and the initial-boundary value problem in bounded domains or external flows. Applications refer generally to non-equilibrium thermodynamics or fluid dynamics applications for molecular flows.

3.3   SOME GENERALIZED MODELS

The analytic and computational complexity of the Boltzmann equation has suggested the derivation of models which, at least in principle, would be supposed to reduce the said complexity. On the other hand, some models, for instance the Enskog equation [5], have been proposed to account for physical features that are not included in the original model.

This section provides a concise presentation focusing on modeling issues that can possibly inspire the derivation of models of crowd dynamics. In more details, the following class of models are treated in the next subsections:

– the so-called BGK model, where the complexity of the right-hand-side term modeling collisions is simplified by replacing the original term by an algebraic structure modeling a trend toward equilibrium;

– the discrete Boltzmann equation [9], where particles are supposed to attain only a finite number of velocities and hybrid models with continuous velocity directions, but finite number of speeds [6];

– the Vlasov equation [7], where interactions are not local, but smeared in the phase space, namely velocity and space; and

– the Enskog equation [5] which includes some dense gas features, for instance finite dimension of the interacting particles and a density dependent interaction rate.

None of these models has a direct influence on the derivation of models of crowd dynamics, but some reasonings underlying their derivation can be used to understand the complex dynamics of human crowds. This specific topic is treated in the critical analysis of the last section which proposes some speculations toward a bridge between the classical kinetic theory and the theory of active particles as well as on the derivation of specific models based on this theory.

3.3.1   THE BGK MODEL

This model has been proposed to reduce the analytic and computational complexity of the right-hand-side term which models, in the Boltzmann equation, interactions between particles. The derivation of this model takes into account the following.

• If the distribution function f is known, then the local Maxwellian can be computed by the first three moments. Formally: $f_e = f_e[f] =: f_e(\rho, \mathbf{U}, \Theta)$, namely density ρ, drift velocity $\mathbf{U}$, and energy Θ.

• A molecular fluid is supposed to have a natural trend to local equilibrium f_e.

• The collision right-hand-side integral operator by a decay term modeling trend to equilibrium.

Consequently, in the absence of an external force field, the model writes as follows:

$$\partial_t f + \mathbf{v} \cdot \nabla_{\mathbf{x}} f = c[f](f_e[f] - f), \tag{3.17}$$

where c gives the decay rate. The latter can be assumed to be a function of the local density ρ, i.e., $c = c(\rho)$. Otherwise, the simplest approximation suggests to take a constant value $c = c_0$. However, the model is nonlinear even in the case of $c = c_0$. In fact f_e nonlinearly depends on f.

Model (3.17) definitely requires less sophisticated tools toward computing than the original Boltzmann equation. However, the Maxwellian distribution is, in the original model, an emerging behavior induced by the dynamics at the microscopic scale, while it is artificially imposed in the BGK model.

3.3.2   THE DISCRETE BOLTZMANN EQUATION AND HYBRID MODELS

Discrete velocity models of the Boltzmann equation can be obtained assuming that particles are allowed to move with a finite number of velocities. The model is an evolution equation for the number densities N_i linked to the admissible velocities $\mathbf{v}_i$ with $i \in L = \{1, \ldots, n\}$. The set $\{N_i(t, \mathbf{x})\}_{i=1}^n$ corresponds, for certain aspects, to the one-particle distribution function of the continuous Boltzmann equation.

The mathematical theory of the discrete kinetic theory is exhaustively treated in the Lecture Notes by Gatignol [9], which provides a detailed analysis of the relevant aspects of the discrete kinetic theory: modeling, analysis of thermodynamic equilibrium, and application to fluid-dynamic problems. The contents mainly refer to a simple mono-atomic gas and to the related thermodynamic aspects. Additional sources of information is the review paper by Platkowski and Illner [13].

The formal expression of the evolution equation corresponds, as for the full Boltzmann equation, to the balance of particles in the elementary volume of the space of the microscopic states:

$$(\partial_t + \mathbf{v}_i \cdot \nabla_{\mathbf{x}}) N_i = \mathcal{N}_i[N_i, N_i] = \frac{1}{2} \sum_{j,h,k=1}^n A_{ij}^{hk} (N_h N_k - N_i N_j), \tag{3.18}$$

which is a system of partial differential equations, where the dependent variables are the number densities linked to the discrete velocities:

$$N_i = N_i(t, \mathbf{x}) : [0, T] \times \mathbb{R}^\nu \to \mathbb{R}_+, \quad i = 1, \ldots, n, \quad \nu = 1, 2, 3.$$

Collisions $(\mathbf{v}_i, \mathbf{v}_j) \longleftrightarrow (\mathbf{v}_h, \mathbf{v}_k)$, are binary, reversible, and preserve mass momentum and energy. Their modeling is left to the so-called *transition rates* A_{ij}^{hk}, which are positive constants and, according to the indistinguishability reversibility properties, satisfy the following relations: $A_{ji}^{hk} = A_{ij}^{kh} = A_{ji}^{kh}$.

Analogously to the Boltzmann equation, it is possible to define the space of collision invariants and of the Maxwellian state [9, 13].

A technical variant to the discrete Boltzmann equation is given by hybrid models, where this definition is used to denote either models with continuous velocity directions, but with discrete speeds and models with discrete velocity directions, but with continuous speed see [6, 9]. Both types of models can offer a useful framework toward crowds modeling, where the number of walkers is not sufficiently large to justify continuity assumptions.

3.3.3   ON THE VLASOV MEAN-FIELD APPROACH

The dynamics in the Boltzmann equation depend only on external actions and short-range interactions. On the other hand, various physical systems are such that also long-range interactions may be significant. The pioneer model, where this specific feature, which is also viewed as a mean-field approximation of interaction, has been introduced is the *Vlasov* equation.

This type of interactions appears in the case of human crowds, where individuals can communicate at distance. Therefore, a brief introduction to mean field models is given.

Let us consider the action $P = P(\mathbf{x}, \mathbf{v}, \mathbf{x}_*, \mathbf{v}_*)$ on the particle with microscopic state $\mathbf{x}, \mathbf{v}$ (test particle) due to the field particles with microscopic state $\mathbf{x}_*, \mathbf{v}_*$. The resultant action is:

$$F[f](t, \mathbf{x}, \mathbf{v}) = \int_{\mathbb{R}^3 \times D_\Omega} P(\mathbf{x}, \mathbf{v}, \mathbf{x}_*, \mathbf{v}_*) \, f(t, \mathbf{x}_*, \mathbf{v}_*) \, d\mathbf{x}_* \, d\mathbf{v}_*, \qquad (3.19)$$

where D_Ω is the domain around the test particle in which the action of the field particle is effectively felt, namely the action P decays with the distance between test and field particles and is equal to zero on the boundary of D_Ω.

Based on the aforesaid assumptions, the mean field equation writes:

$$\partial_t f + (\mathbf{v} \cdot \nabla_\mathbf{x}) f + \mathbf{F} \cdot \nabla_\mathbf{v} f + \nabla_\mathbf{v} \cdot (F[f] \, f) = 0, \qquad (3.20)$$

where $\mathbf{F}$ is the positional macroscopic force acting on the system.

3.3.4   ON THE ENSKOG EQUATION APPROACH

The *Enskog* equation introduces some effects of the finite dimensions of the particles which appear in two ways.

1. The distribution functions of the interacting pairs (indeed, only binary collisions are considered) are computed in the centers of the two spheres, namely not in a common point for both distributions.

2. The collision frequency is reduced by the dimension of the interacting spheres, which reduces the probability of further interactions by shielding the free volume available for further interactions. This amounts to introducing a functional of the local density which

correlate the distribution functions of the interacting pairs and increases with increasing local density.

This model has been interpreted for a long time as the first step toward the derivation of kinetic models for dense fluids. On the other hand, the limitation to binary mixtures technically prevents such interesting generalization. In fact, multiple interactions appear to be necessary to describe the physics of transition from rarefied to dense fluid.

However, as it is, the model offers to applied mathematicians a variety of challenging problems such as existence and uniqueness of solutions to mathematical problems and asymptotic limits either to hydrodynamics when the intermolecular distances tends to zero, or to the Boltzmann equation when the dimension of the particles is allowed to tend to zero. The book [5] was devoted to this model and mainly to the analytic problems generated by its applications. The bibliography reported in it also includes useful indications in the field of physics.

The mathematical structure of this model is not reported here as it is a technical modification of the Boltzmann equation, where nonlocal interactions are accounted for by modeling particles as balls with constant radius, while the interaction frequency is modified in order to account for the finite dimension of the particles, namely it is a functional of the local density which increases as the density decreases.

3.4   ON THE KINETIC THEORY OF ACTIVE PARTICLES

This section refers to the derivation of mathematical structures to provide the conceptual framework for the derivation of models suitable to describe the dynamics of human crowds either in an unbounded domain in $\mathbb{R}^2$ with walkers who move along one or more prefixed directions and to crowds confined in a two-dimensional bounded domain Σ, with walls and one or more exit doors, which can also include internal obstacles. Exit doors might either correspond evacuation to an unbounded domain or moving into a new walking area with different geometrical and quality features. If the crowd moves in an unbounded domain, Σ_0 denotes the domain initially occupied by the crowd.

As we have seen in Chapter 2, it is useful using dimensionless quantities by referring linear coordinates to ℓ which is selected either by the diameter of the circular domain containing or, in the case of an unbounded domain, by the diameter of the circular domain containing the domain initially occupied by the crowd, i.e., Σ_0. The speed v, namely the velocity modulus of the velocity $\mathbf{v}$, is referred to a limit velocity v_ℓ defined, as in Chapter 2, by the limit speed of fast walkers in a venue dependent under free flow conditions. Dimensionless time, denoted by t, is obtained by dividing the real time by $T = \ell/v_\ell$. Often, it is useful using polar coordinates $\mathbf{v} = \{v\cos\theta, v\sin\theta\}$, with obvious meaning of notations.

Let us now define, repeating some concepts already delivered in Chapter 2, the minimal variables and parameters that can characterize the aforementioned mathematical structures.

- Walkers have a visibility zone $\Omega_v = [\theta - \Theta, \theta + \Theta] \times [0, R]$, where Θ and R define, respectively, the visual angle and distance.

- The parameter $\alpha \in [0, 1]$ denotes the quality of the venue, where the dynamics occur.

- The variable $u \in [0, 1]$ denotes a specific emotional state which can be defined within the framework of each specific case study under consideration. If the said emotional state is shared by all individuals in the crowd, a parameter $\beta \in [0, 1]$ replaces the variable u.

- If polar coordinates are used, the distribution function $f = f(t, \mathbf{x}, v, \theta, u)$ is used instead of $f = f(t, \mathbf{x}, \mathbf{v}, u)$, where the emotional state u has been inserted into the microscopic state which is now $\mathbf{x}, v, \theta, u$.

- The distribution function f refers to the packing density ρ_M so that, if f is locally integrable, then $f(t, \mathbf{x}, \mathbf{v}, u)\, d\mathbf{x}\, d\mathbf{v}\, du$ denotes the dimensionless density of interacting particles which, at time t, are in the phase elementary domain $[\mathbf{x}, \mathbf{x} + d\mathbf{x}] \times [\mathbf{v}, \mathbf{v} + d\mathbf{v}] \times [u, u + du]$.

Macroscopic quantities, for example local density, total number of walkers, local dimensionless mean velocity and flow, can be obtained by the same type of calculations used in the case of classical particles, accounting for integration over space and velocity in two dimensions as well as of the additional integration over the activity variables.

Let us first consider a crowd not yet subdivided into functional subsystems and show how macroscopic quantities can be computed. In detail, the local density $\rho = \rho(t, \mathbf{x})$ referred to ρ_M, is given by

$$\rho(t, \mathbf{x}) = \int_0^1 \int_0^{2\pi} \int_0^1 f(t, \mathbf{x}, v, \theta, u)\, v\, dv\, d\theta\, du, \tag{3.21}$$

where $v\, d\theta\, dv = d\mathbf{v}$, while the flow $\mathbf{Q} = \mathbf{Q}(t, \mathbf{x})$ is computed as follows:

$$\mathbf{Q}(t, \mathbf{x}) = \int_0^1 \int_{D_v} \mathbf{v}\, f(t, \mathbf{x}, \mathbf{v}, u)\, d\mathbf{v}\, du$$

$$= \int_0^1 \int_0^{2\pi} \int_0^1 (v \cos\theta\, \mathbf{i} + v \sin\theta\, \mathbf{j})\, f(t, \mathbf{x}, v, \theta, u)\, v\, dv\, d\theta\, du, \tag{3.22}$$

where $\mathbf{i}$ and $\mathbf{j}$ are unit vectors of the orthogonal cartesian frame. Calculation of the mean velocity follows:

$$\boldsymbol{\xi}(t, \mathbf{x}) = \mathbf{Q}(t, \mathbf{x})\, \rho(t, \mathbf{x}). \tag{3.23}$$

Analogous calculations can be developed for a quantity corresponding to the mean value of the speed somehow related to a local kinetic energy in the one-directional motion:

$$\mathcal{E}(t, \mathbf{x}) = \frac{1}{2\, \rho(t, \mathbf{x})} \int_0^1 \int_{D_v} v^2 f(t, \mathbf{x}, \mathbf{v}, u)\, d\mathbf{v}\, du, \tag{3.24}$$

where, however, the concept of individual mass cannot be deterministically clarified.

The derivation of a mathematical structure, deemed to provide a general framework for the derivation of models, can be developed referring to [4]. Hence, interactions involve the following particles.

- *Candidate particles* with probability distribution $f(t, \mathbf{x}, \mathbf{v}_*, u_*)$ which are deemed, by interaction with the field particles which are in their visibility domain, to acquire the state of the test particle.

- *Field particles* with probability distribution $f(t, \mathbf{x}^*, \mathbf{v}^*, u^*)$ which lose their state, by interaction with the candidate particles, where interactions occur for all particles in the visibility domain $\mathbf{x}^* \in \Omega_{\mathbf{v}}(\mathbf{x})$.

- *Test particle* with probability distribution $f(t, \mathbf{x}, \mathbf{v}, u)$, representative of the whole system, which gains due to transition into their state of candidate particles due to interaction with the field particles, while test particles lose their state, by interaction with the field particles.

Remark 3.2   The role in the interactions, which has been defined above, is a statistical concept as particles are not deterministically identified.

In general, the modeling of interactions can be described by the following quantities.

- *The encounter rate*

$$\eta(f; \mathbf{x}, \mathbf{x}^*, \mathbf{v}_*, \mathbf{v}^*, u_*, u^*, \alpha)$$

which denotes the interaction frequency between candidate and field particles, while the argument changes for the other type of interactions.

- *Transition probability density*

$$\mathcal{A}(\mathbf{v}_* \to \mathbf{v} | f; \mathbf{x}, \mathbf{x}^*, \mathbf{v}_*, \mathbf{v}^*, u_*, u^*, \alpha)$$

whose integral over all outputs is equal to one.

Remark 3.3   Both $\Omega_{\mathbf{v}}$ and v_ℓ should be related to the shape and quality of the venue where the dynamics occurs. The geometrical shape, for instance the presence of walls, can reduce both visibility angle and distance.

Mathematical structures, suitable for providing conceptual basis for derivation of models, can be obtained by equating the transport of f to the net flow in the elementary volume of the space of microscopic states as it is induced by interactions. A quite general structure is as follows:

$$\partial_t f(t, \mathbf{x}, \mathbf{v}, u) + \mathbf{v} \cdot \nabla_{\mathbf{x}} f(t, \mathbf{x}, \mathbf{v}, u)$$
$$= \int_0^1 \int_0^1 \int_{D_v} \int_{D_v} \int_{\Omega_{v_*}(\mathbf{x})} \eta(f; \mathbf{x}, \mathbf{x}^*, \mathbf{v}_*, \mathbf{v}^*, u_*, u^*, \alpha)$$
$$\mathcal{A}(\mathbf{v}_* \to \mathbf{v} | f; \mathbf{x}, \mathbf{x}^*, \mathbf{v}_*, \mathbf{v}^*, u_*, u^*, \alpha) \times f(t, \mathbf{x}, \mathbf{v}_*, u_*)$$
$$f(t, \mathbf{x}^*, \mathbf{v}^*, u^*) \, d\mathbf{x}^* \, d\mathbf{v}_* \, d\mathbf{v}^* \, du_* du^*$$
$$- f(t, \mathbf{x}, \mathbf{v}, u) \int_0^1 \int_{D_v} \int_{\Omega_v(\mathbf{x})} \eta(f; \mathbf{x}, \mathbf{x}^*, \mathbf{v}_*, \mathbf{v}^*, u_*, u^*, \alpha)$$
$$f(t, \mathbf{x}^*, \mathbf{v}^*, u^*) \, d\mathbf{x}^* \, d\mathbf{v}^* \, du^*. \tag{3.25}$$

In general, walkers can be subdivided into a number n of functional subsystems corresponding to different types of walking directions or exits in case of evacuation, then the probability distribution is labeled by the subscript $i = 1, \ldots, n$. Technical calculations yield:

$$\partial_t f_i(t, \mathbf{x}, \mathbf{v}, u) + \mathbf{v} \cdot \nabla_{\mathbf{x}} f_i(t, \mathbf{x}, \mathbf{v}, u)$$
$$= \sum_{k=1}^n \int_0^1 \int_0^1 \int_{D_v} \int_{D_v} \int_{\Omega_{v_*}(\mathbf{x})} \eta_{ik}(\mathbf{f}; \mathbf{x}, \mathbf{x}^*, \mathbf{v}_*, \mathbf{v}^*, u_*, u^*, \alpha)$$
$$\mathcal{A}_{ik}(\mathbf{v}_* \to \mathbf{v} | \mathbf{f}; \mathbf{x}, \mathbf{x}^*, \mathbf{v}_*, \mathbf{v}^*, u_*, u^*, \alpha) \times f_i(t, \mathbf{x}, \mathbf{v}_*, u_*)$$
$$f_k(t, \mathbf{x}, \mathbf{v}^*, u^*) \, d\mathbf{x}^* \, d\mathbf{v}_* \, d\mathbf{v}^* \, du_* du^*$$
$$- f_i(t, \mathbf{x}, \mathbf{v}, u) \sum_{k=1}^n \int_0^1 \int_{D_v} \int_{\Omega_v(\mathbf{x})} \eta_{ik}(\mathbf{f}; \mathbf{x}, \mathbf{x}^*, \mathbf{v}_*, \mathbf{v}^*, u_*, u^*, \alpha)$$
$$f_k(t, \mathbf{x}, \mathbf{v}^*, u^*) \, d\mathbf{x}^* \, d\mathbf{v}^* \, du^*, \tag{3.26}$$

where the encounter rate η_{ik} and the transition probability density $\mathcal{A}_{ik}$ accounts for interactions with the same functional subsystem (FS) and across functional subsystems (FSs).

The mathematical structures (3.25) and (3.26) provide the general framework which can be further simplified or generalized before the derivation of models. Some technical developments can be rapidly indicated to be further clarified in the specific models proposed in the next chapter. In more detail, the following ones are brought to the attention of the reader.

1. Semi-discrete (hybrid) correspond to the assumption that the velocity directions can only attain a finite number of finite values, while the speed is continuous along each direction.

2. Discrete models correspond to the assumption that both speed and velocity directions can only attain a finite number of finite values.

3. First-order models are obtained assuming that interactions modify only the velocity directions, while the speed is delivered by phenomenological models that refer it to the local density and density gradients.

4. Quantities which play a role in the dynamics of interactions are estimated by a weighted average within the visibility domain.

Remark 3.4   As mentioned, a variety of models have made use of discrete velocity models by the basic assumption that the velocity can attain only a finite number of directions and speeds, respectively,

$$\{\theta_1 = 0, \ldots, \theta_h, \ldots, \theta_H = 2\pi/(H-1)\} \qquad \text{and} \qquad \{v_1 = 0, \ldots, v_k, \ldots, v_K\},$$

where H and K denote, respectively, the number of directions and of speeds.

This assumption rather than a simplification should be seen as a possible way of relaxing the continuity assumption of the probability distribution over the micro-state.

Let us now consider the mathematical structures which can be generated according to Remark 3.4 in the case of one functional subsystem only, where the activity variable is a parameter β equally shared by all individuals in the crowd. Then the state of the system is described by the discrete probability:

$$\mathbf{f} = \{f_{hk}\}, \qquad f_{hk}(t, \mathbf{x}) = f(t, \mathbf{x}, \theta_h, v_k), \qquad h = 1, \ldots, H, \qquad k = 1, \ldots, K. \tag{3.27}$$

The terms modeling interactions are follows.

- *The discrete encounter rate* $\eta_{rs}^{pq}(f_{rs}, f_{pq}|\mathbf{x}, \mathbf{x}^*, \alpha, \beta)$ which denotes the interaction frequency between rs-candidate and pq-field particles.

- *The discrete transition probability density* $\mathcal{A}(rs \to hk|f_{rs}, f_{pq}, \mathbf{x}, \mathbf{x}^*, \alpha, \beta)$, whose integral over all outputs is equal to one.

Calculations analogous to those we have seen above yields

$$\partial_t f_{hk}(t, \mathbf{x}) + \mathbf{v}_{hk} \cdot \nabla_{\mathbf{x}} f_{hk}(t, \mathbf{x})$$
$$= \sum_{rs} \sum_{pq} \int_{\Omega_v(\mathbf{x})} \eta_{rs}^{pq}(f_{rs}, f_{pq}|\mathbf{x}, \mathbf{x}^*, \alpha, \beta)$$
$$\mathcal{A}(rs \to hk|f_{rs}, f_{pq}, \mathbf{x}, \mathbf{x}^*, \alpha, \beta) f_{rs}(t, \mathbf{x}) f_{pq}(t, \mathbf{x}^*)$$
$$- f_{hk}(t, \mathbf{x}) \sum_{pq} \int_{\Omega_v(\mathbf{x})} \eta_{hk}^{pq}(f_{rs}, f_{pq}|\mathbf{x}, \mathbf{x}^*, \alpha, \beta) f_{pq}(t, \mathbf{x}^*). \tag{3.28}$$

In addition, one might rapidly consider an additional subdivision into FSs, for instance FSs corresponding to different types of walking abilities, models where the emotional state can attain different values and various others. However, we limit our presentation to the above structures, namely to those which have been effectively used in the literature. The models that will be presented in the next chapter should enlighten how different features can be inserted in the structures we have proposed in this chapter.

3.5   CRITICAL ANALYSIS

It has been mentioned various times in this chapter that substantial differences distinguish the two theories related, respectively, to classical and active particles. A detailed analysis of these differences can contribute both to understand how the theory for active particles developed to model various systems of interacting living particles, and to contribute to a strategy to derive specific models. In addition, it is important to stress that the active particle theory is not a straightforward generalization of the Boltzmann equation.

This last remark applies not only to models of crowd dynamics but also to the pioneer works by Prigogine and Herman [14] and to the sharp revisiting by Paveri Fontana [12]. We will return to this matter later after having brought to the reader's attention the following conceptual differences which have been selected, without claiming completeness, in view of the applications to crowd dynamics.

- *Multiple and nonlocal interactions:* The derivation of the Boltzmann equation is based on the assumption that only binary interactions are taken into account. Then the physical validity of the equation is limited to a diluted gas. On the other hand, models of crowds need to describe also high density conditions which might have an important role in managing crisis situations. This matter is tackled by accounting for nonlocal interactions as each particle interacts with all particles in the visibility/sensitivity domain as well as the modeling of repulsive forces depending on the distance between interacting walkers. This feature generates non trivial computational problems, but it tackles, at least partially, high density features.

- *Nonlinearly additive interactions and equilibrium:* Interactions in classical kinetic theory are assumed to be reversible and preserve not only mass, but also momentum and energy. Some generalized models also account for dissipative interactions and chemically reactive gases. The conservative case leads to define Maxwellian equilibrium conditions as we have seen in Section 3.2. This is not the case of interactions in a crowd, where individual energy can be delivered or dissipated. On the other hand, empirical data on the so-called steady uniform and one directional flow shows velocity diagrams that indicate how the mean velocity decays with the density.

- *Venue-dependent dynamics:* The dynamics of self-propelled particles depend on the quality of the venue which can reduce, when it decays, the speed and interaction rate between active particles. This feature is not accounted in the classical kinetic theory, where the interaction rate is mechanical depending on the relative velocity and on the cross-sectional area.

- *Continuity problems of the probability distributions:* The assumption of continuity of the probability distribution is justified in the case of the Boltzmann equation as the number of particles is sufficiently high to motivate this assumption. On the other hand, the number of walkers in a crowd makes less appropriate this assumption which has been criticized already in the case of vehicular traffic [8]. This drawback can be partially accounted for by discrete velocity models, where the dynamics involves groups of particles in the elementary volume of the phase space.

- *Modeling mixtures:* This has been developed in classical kinetic theory to model mixtures of different particles which might even generate chemical reactions. Mixtures in crowd models have to be considered a necessary feature related to different roles that individual entities can have in the crowd, for instance leaders, rescue teams, antagonist groups, etc.

Some conceptual differences have been critically analyzed in the above items which enlighten some, definitely not all, differences. This is not surprising as the physics of the inert matter is highly different from the physics of living entities. Henceforth, a different mathematics follows.

The physics of vehicular traffic, which has been enlightened in the book by Kerner [10], shows some similarity with that of crowds, but it also shows substantial differences. However, physicists and mathematicians have argued on the difference between the Boltzmann equation and kinetic models of vehicular traffic. Indeed, substantial differences distinguish the two approaches. However, rather than stressing this concept we simply report what has been observed in Chapter 1 of [2]:

The interpretation that Prigogine's model is a technical modification of the Boltzmann equation is definitely unfair, as his book [14] contains new general ideas on the modeling of interactions between vehicles by tools of probability theory and on the modeling of heterogeneous behaviors. In fact, this book anticipates issues that now are developed in various fields of applications such as the dynamics of crowds and swarms. The great conceptual novelty is that the overall state of a large population is described by a probability distribution over the micro-scale state of interacting entities and that the output of interactions is delivered by probability rules rather than by the deterministic causality principle.

Bearing all of the above in mind, let us now look ahead to the derivation of models, by the kinetic theory approach, which essentially consists first in selecting the mathematical structure appropriate to account for the specific features of each case study under consideration, second by modeling all interactions involving individuals in the crowd, and last by inserting them into the said structure. The next chapter will show how this strategy can be applied in the derivation of crowd models.

3.6 BIBLIOGRAPHY

[1] N. Bellomo, Ed., *Lecture Notes on the Mathematical Theory of the Boltzmann Equation*, World Scientific, Singapore, 1995. 35

[2] N. Bellomo, A. Bellouquid, L. Gibelli, and N. Outada, *A Quest Towards a Mathematical Theory of Living Systems*, Birkhäuser-Springer, New York, 2017. DOI: 10.1007/978-3-319-57436-3 34, 38, 49

[3] N. Bellomo, A. Bellouquid, and D. Knopoff, From the micro-scale to collective crowd dynamics, *Multiscale Modelling and Simulation*, 11:943–963, 2013. DOI: 10.1137/130904569 33

[4] N. Bellomo and L. Gibelli, Toward a behavioral-social dynamics of pedestrian crowds, *Mathematical Models and Methods in Applied Sciences*, 25:2417–2437, 2015. DOI: 10.1142/S0218202515400138 45

[5] N. Bellomo, M. Lachowicz, J. Polewczak, and G. Toscani, *The Enskog Equation*, World Scientific, Singapore, 1991. 40, 43

[6] H. Cabannes, L. Pasol, and K. G. Roesner, Study of a new semi-continuous model of the Boltzmann equation, *European Journal of Mechanics, B/Fluids*, 21:751–760, 2002. DOI: 10.1016/s0997-7546(02)01214-1 40, 42

[7] C. Cercignani, R. Illner, and M. Pulvirenti, *The Mathematical Theory of Diluted Gas*, Springer, Heidelberg, New York, 1993. DOI: 10.1007/978-1-4419-8524-8 34, 36, 38, 39, 40

[8] C. Daganzo, Requem for second-order fluid approximation of traffic flow, *Transportation Research B*, 29:277–286, 1996. DOI: 10.1016/0191-2615(95)00007-Z 49

[9] R. Gatignol, Theorie cinétique des gaz a repartition discréte de vitesses, *Lectures Notes in Physics*, 36, Springer, 1975. DOI: 10.1007/3-540-07156-3 40, 41, 42

[10] B. S. Kerner, *The Physics of Traffic* Springer, Heidelberg, New York, 2004. DOI: 10.1007/978-3-540-40986-1 49

[11] D. Kim and A. Quaini, A kinetic theory approach to model pedestrian dynamics in bounded domains with obstacles, *Kinetic and Related Models*, 12(6):1273–1296, 2019. DOI: 10.3934/krm.2019049 33

[12] S. Paveri Fontana, On Boltzmann like treatments for traffic flow, *Transportation Research*, 9:225–235, 1975. DOI: 10.1016/0041-1647(75)90063-5 48

[13] T. Platkowsky and R. Illner, Discrete velocity models of the Boltzmann equation: Ā survey on the mathematical aspects of the theory, *SIAM Review*, 30:213–255, 1988. DOI: 10.1137/1030045 41, 42

[14] I. Prigogine and R. Herman, *Kinetic Theory of Vehicular Traffic*, Elsevier, New York, 1971. 48, 49

CHAPTER 4

Kinetic Theory Models and Applications

Abstract This chapter shows how to derive models of crowd dynamics from the mathematical structure presented in Chapter 3 and discusses the numerical approaches available for their solutions. First, the most recent developments of kinetic theory modeling of human crowds are reviewed with particular attention to literature contributions that account for pedestrians' emotional states and deal with contagion dynamics. Next, numerical methods for solving kinetic theory models are discussed, including the Monte Carlo particle methods that turn out to be particularly suitable for mitigating the computational cost of simulations. Finally, a specific crowd dynamics model is described, and sample simulation results are discussed.

4.1   INTRODUCTION

This chapter is devoted to modeling and applications with the objective of showing how the derivation of models from the mathematical structures presented in Section 3.4 can be developed toward applications. As previously mentioned, the mathematical description of how each pedestrian interacts with other pedestrians within her/his visibility domain is the key ingredient. Unlike collisions between classical particles, interactions between walkers are non-reversible, non-local, and depend on the emotional state of the people in the crowd. The quality of the environment where pedestrians moves is an additional important aspect that needs to be accounted for in the modeling process.

In this chapter emphasis is put on the influence of pedestrians emotional states on the collective dynamics. A high level of stress in crowds may be triggered by various reasons, ranging from the anxiety to reach the exit during emergency evacuations to overcrowding conditions that may occur in rush hours. The models presented in the following sections assume that the emotional states of pedestrians do not depend on time, i.e., they are not induced and/or propagated by individual communications via vocal or visual signaling. However, in principle, emotional states spread out and this dynamic significantly affects the overall crowd behavior. This latter aspect, which has been defined *contagion* in [53], will be treated in the next chapter.

Sample simulations based on a Monte Carlo particle method are reported to show how models can be validated and their predictive ability can be assessed. This method of solution has been chosen because it is ideally suited to deal with the sophisticated pedestrians interactions, designed through stochastic game theory, which kinetic models are based on.

More specifically, the contents of the chapter are presented as follows. Section 4.2 provides a concise review of the more recent developments of the kinetic theory approach to crowd modeling. The main features of crowd models are described and referred to as the mathematical structures presented in Chapter 3. Section 4.3 outlines the computational tools which can be used to numerically solve the kinetic theory models. Among them, Monte Carlo particle methods are the most appropriate and, therefore, their algorithmic structure is described in some more detail. Section 4.4 selects a well-defined model, characterized by emotional states equally shared by all pedestrians. It is shown how the model can be derived by implementing the behavioral strategy of walkers into the general mathematical structure. Section 4.5 presents and critically analyzes some sample simulations reported in the literature. In particular, two aspects are specifically considered, namely the ability of kinetic models to depict the segregation phenomena that occurs in pedestrians counter flows and to provide a reasonable description of pedestrians behavior during emergency evacuation. Section 4.6 provides a critical analysis focused on the contents of this chapter and discusses some interesting lines of future research in this area.

4.2   A SURVEY OF KINETIC THEORY MODELS

This section concisely reviews the models derived by the kinetic theory approach. The interested reader can find further information in the recent survey [2] devoted to modeling of traffic, crowds, and swarms by means of kinetic theory and swarms methods.

Mathematical models of crowd dynamics have been initially proposed referring to the mathematical structure corresponding to discrete velocity directions delivered by Eq. (3.28). This modeling approach was developed in [10] under the assumption that walkers move at the same speed while their velocity direction can only take discrete values uniformly distributed over the unit circle. Furthermore, it is assumed that, due to the interactions, the candidate walker changes velocity direction based on the weighted selection of three stimuli, namely the desire of following a prescribed local direction, the attempt to avoid overcrowded areas and the tendency to do what all other walkers do, which is often referred to as "herding behavior." Two weighting parameters have been introduced, namely the local density and the level of stress of walkers. More specifically, the larger is the local density the greater is the desire of avoiding overcrowded areas. Likewise, the larger is the level of stress the stronger is the attraction toward the main stream.

As a further development, an hybrid model has been proposed in [1] and subsequently applied to study the evacuation dynamics in [48]. In these works, the velocity directions are still discrete while the speed takes continuous values in the domain [0, 1] base on the *velocity diagram* as given by empirical data [47–49]. The latter provides the relationship between the mean velocity vs. density and has been object of many thoroughly studies over the years [48,

48, 49]. Additional bibliography enlightened by sharp reasoning on the use of empirical data toward design and tuning of models is reported in [46].

Discrete velocity models have been applied in [22] and [34] to compute the evacuation time from a venue that includes internal obstacles. The authors show how the qualitative analysis, proposed in [10] for the solutions to the initial value problem in unbounded domains, can be technically generalized to include interactions with boundaries.

An analytic study has been carried out in [8] to derive a macroscopic description of crowds from kinetic theory models. This approach refers the derivation of macroscopic models to the dynamics at the microscopic, individual-based scale rather than by conservation equations closed by heuristic models for the constitutive behavior.

The approach to modeling crowds by means of a continuous velocity variable has been proposed in [12] and further developed in [30] where the validation of crowd models has also been addressed. The basic idea is that models should reproduce quantitatively empirical data which are generally obtained in steady uniform flow, and qualitatively collective emerging behaviors which are repetitively observed in specific physical conditions. In this respect, the model proposed in [30] was shown to depict the self-organized formation of lanes of uniform walking directions in pedestrian counter flows and a velocity diagram consistent with empirical data. This latter has also been obtained in [24] for a simple discrete velocity model of vehicular traffic. A key feature of both models is that the relationship obtained between density and velocity was not artificially inserted into the modeling but it is an emerging property of the underlying microscopic dynamics.

Note that all models based on either discrete or continuous velocities require a detailed description of the interactions at the microscopic scale, namely, the various strategies expressed by walkers to select their trajectories and react to the presence of other walkers. Interesting contributions have been given by [39], where the authors use empirical data delivered by experiments specifically designed to achieve the aforementioned information. Granular behaviors and contact problems have been studied in [23, 37], while the implementation of the dynamics at the microscopic scale into possible hierarchies of models has been developed in [19, 20].

Note that empirical date play a crucial role in modeling of interactions between pedestrians. The research contribution delivered in [17, 18] proposes a new vision on the future development of the experimental research activity.

The very brief review of this section has been limited to models of behavioral crowds, but not yet to models that account for the propagation of emotional states. This topic has been introduced in some recent papers [14, 15] and it will be discussed in the next chapter.

4.3 COMPUTATIONAL METHODS

Kinetic equations are nonlinear integro-differential equations in which the unknown function depends on a large number of variables. The complicated mathematical structure makes their analytical solution unfeasible and, therefore, one has to resort to the numerical approach, especially for studying cases of practical interest. The most widely used strategy consists in decoupling the transport and collision terms by time-splitting the evolution operator into a *drift step*, in which collisions are neglected, and a *collision step*, in which there is no spatial motion.

Numerical methods can be roughly grouped into three categories depending on how drift and collision steps are dealt with, namely regular, semi-regular and particle. Valuable references are provided by standard books [4, 16, 44] as well as recent review papers [2, 21].

In both regular and semi-regular methods the distribution function is discretized on a structured and/or unconstructed grid in the phase space [4]. Accordingly, the drift step amounts of solving a system of hyperbolic conservation laws coupled at the boundaries. Their discretization can be done by a variety of ways, including finite-difference, finite-volume, finite-element, and spectral methods [25]. The collision step amounts of solving a spatially homogeneous relaxation equation. This is the more computationally demanding part since it involves the computation of the high-dimensional integral which defines the collision operator. Regular and semi-regular methods adopt different strategies to evaluate the collision term.

Most of the regular methods use a Galerkin discretization of the velocity space [21]. According to these methods, the velocity dependence of the distribution function is expanded in a set of trial functions, being the expansion coefficients depending on position and time. The Galerkin ansatz is then substituted in the space homogeneous relaxation equation which is subsequently multiplied by test functions and integrated in the velocity space. Note that in the Galerkin approach, test functions are taken the same as trial functions. The above procedure leads to a coupled system of ordinary differential equations equations for the expansion coefficients.

Galerkin discretization can be further distinguished depending on the employed basis functions. In the Fourier–Galerkin approach, the distribution function is expanded in trigonometric polynomials and the fast Fourier Transform is used to accelerate the computation of the collision integral in the velocity space [26, 28, 38, 43], while discontinuous Galerkin methods adopt discontinuous piecewise polynomials as test and trial functions [3, 29].

Hybrid approaches have been also implemented where the distribution function is expanded in Laguerre polynomials with respect to the velocity components parallel to solid surfaces while quadratic finite-element functions have been used for the normal velocity component [42, 50].

In semi-regular methods, the collision integral is computed by Monte Carlo or quasi-Monte Carlo quadrature. These schemes originate from the work by Nordsieck and Hicks [41] and have been further developed over the years by a number of authors [5, 27].

Particle methods originate from the Direct Simulation Monte Carlo (DSMC) scheme [16]. This latter has been originally introduced based on physical reasoning and it has been later proved to converge, in a suitable limit, to the solution of the Boltzmann equation [52]. The distribution function is represented by a number of computational particles which move in the simulation domain and collide according to stochastic rules derived from the kinetic equations.

The simulation domain is covered by a mesh of cells which are used for collecting particles that may collide and sampling the macroscopic fields. Macroscopic flow properties are obtained by time-averaging particle properties. Several variants of the DSMC have been proposed over the years which differ in the way the collision step is performed. These methods include the majoring Frequency scheme [33], null Collision scheme [35], Nanbu scheme [40] and its modified version [7], Ballot Box scheme [54], and Simplified Bernoulli trials scheme [51].

Particle schemes offer a number of attractive features which make them the most popular and widely used simulation methods for solving kinetic equations. Among other things, they permit the easily handling of complex geometries while keeping the computational burden at a reasonable level. Furthermore, they permit easily insertion of new physics and/or dealing with complex interaction rules between particles. By contract, particle methods not well suited to simulate unsteady problems since, in this case, the possibility of time averaging is lost or reduced. Acceptable accuracy can only be achieved by increasing the number of simulation particles or superposing several simulation snapshots obtained from statistically independent simulations. However, in both cases, the computing effort is considerably increased. Weighted particles can also be used to increase the results' accuracy without additional computational cost [45].

Note that the adoption of a grid to discretize the distribution function limits the applicability of regular and semi-regular schemes to problems where particular symmetries reduce the number of independent variables. However, these methods are an appealing alternative to particle methods in studying unsteady flows since it permits one to adopt a direct steady-state formulation which leads to a huge savings in computational cost, at least if the transient behavior is not of interest.

4.4   SELECTION OF A KINETIC MODEL

This section shows how specific models can be derived from the mathematical structure (3.28) by means of a detailed analysis of the interactions at the microscopic scale. The discussion below is restricted to the models developed in [6, 14, 30]. The underlying assumptions are presented, while the interested reader is addressed to the articles cited above for the detailed expressions of mathematical equations which are not repeated here.

Let us consider the simpler case of a crowd composed of only one functional subsystem and let us assume that all walkers share the same emotional state. This latter is quantified by the parameter $\beta \in [0, 1]$, while the quality of the environment where walkers move is defined by the parameter $\alpha \in [0, 1]$. Both parameters have been introduced in Chapter 3. We first consider the modeling in unbounded domains and, subsequently, we discuss how the presence of obstacles and walls modifies the structure of the model.

The basic assumptions of the model are as follows.

1. All a-particles have the visibility angles Θ and $-\Theta$, symmetric with respect to the velocity direction, and a visibility radius R depending on the quality and shape of the venue; Θ and R define the visibility area, denoted by Ω.

2. The decision process is such that walkers first change, in probability, the direction of movement and then modify their speed. Three stimuli are assumed to contribute to the change of the walking direction, namely, the desire to reach a defined target, the tendency to do what others do, and the attempt to avoid overcrowded areas. These stimuli are represented by the three unit vectors $\boldsymbol{\nu}^{(t)}$, $\boldsymbol{\nu}^{(s)}$, and $\boldsymbol{\nu}^{(v)}$, respectively, and the preferred walking direction results from their weighted sum. Note that the weights which enter in the latter depends on the quality of the venue, emotional state, and local density.

3. Walkers moving in a certain direction perceive a density which is higher than the real one in the presence of positive gradients, while it is lower than the real density when the gradient is negative. The larger is the perceived density, the more walkers try to move away from the more congested area taking the direction of $\boldsymbol{\nu}^{(v)}$. By contrast, the lower the perceived density, the more walkers head for the target identified by $\boldsymbol{\nu}^{(t)}$ unless their level of anxiety is high, in case of which they tend to follow the mean pedestrian stream as given by $\boldsymbol{\nu}^{(s)}$.

4. The modification of the speed is such that if the walker's speed is greater than (or equal to) the mean local speed in his/her visibility domain, walkers either maintain their speed or decelerate to a speed which is as much lower as the local density becomes higher. Otherwise, if the walker's speed is lower than the mean speed, walkers either maintain their speed or accelerate to the mean speed which is as much higher as the local density becomes lower.

5. Increasing values of α, which correspond to environments of better quality, leads to higher walking speeds and a more rapid change of direction. Increasing values of β, which corresponds to higher stressful conditions, leads to a stronger tendency of walkers to follow the main pedestrian stream rather than looking for less congested areas.

These assumptions can be made formal and transferred into the aforementioned differential structure as follows.

Step 1. Representation of the system: The crowd is described by the distribution function over the micro-state of walkers. This latter is identified by position and velocity, which is given in polar coordinates:

$$f = f(t, \mathbf{v}|\alpha, \beta) = f(t, \mathbf{x}, v, \theta|\alpha, \beta), \tag{4.1}$$

where

$$\mathbf{v} = \{v, \theta\}, \quad \mathbf{x} \in \Sigma \subset \mathbb{R}^2, \quad v \in [0, 1], \quad \theta \in [0, 2\pi) \quad \alpha, \beta \in [0, 1].$$

Remark 4.1   Macroscopic quantities can be obtained by velocity-weighted moments of the distribution function. In particular, the local *density* and the *mean velocity* are given as follows:

$$\rho(t, \mathbf{x}|\alpha, \beta) = \int_{D_{\mathbf{v}}} f(t, \mathbf{x}, \mathbf{v}|\alpha, \beta) \, d\mathbf{v}, \tag{4.2}$$

and

$$\boldsymbol{\xi}(t, \mathbf{x}|\alpha, \beta) = \frac{1}{\rho_i(t, \mathbf{x})} \int_{D_{\mathbf{v}}} f(t, \mathbf{x}, \mathbf{v}|\alpha, \beta) \, d\mathbf{v}, \tag{4.3}$$

or

$$\rho(t, \mathbf{x}|\alpha, \beta) = \int_0^1 \int_0^{2\pi} f(t, \mathbf{x}, v, \theta|\alpha, \beta) \, v \, dv \, d\theta, \tag{4.4}$$

and

$$\boldsymbol{\xi}(t, \mathbf{x}|\alpha, \beta) = \frac{1}{\rho(t, \mathbf{x})} \int_0^1 \int_0^{2\pi} \mathbf{v} \, f(t, \mathbf{x}, v, \theta|\alpha, \beta) \, v \, dv \, d\theta, \tag{4.5}$$

as $d\mathbf{v} = v \, dv \, d\theta$ and $D_{\mathbf{v}} = [0, 1] \times [0, 2\pi)$.

Step 2. Quantitative modeling of interactions: According to Section 3.4, interactions between walkers are defined by the *interaction rate* η, and the *transition probability density* $\mathcal{A}$, which, in general, depends on the micro-states and on the distribution functions of the interacting particles, as well as on the quality of the environment where walkers move. The interaction rate models the frequency by which a candidate (resp. test) particle in $\mathbf{x}$ interacts, in the visibility domain, with a field particle in $\mathbf{x}^*$. Its formal expression reads:

$$\eta[f](\mathbf{x}, \mathbf{x}^*, \mathbf{v}_*, \mathbf{v}^*; \alpha, \beta), \quad (\text{resp.} \quad \eta[f](\mathbf{x}, \mathbf{x}^*, \mathbf{v}, \mathbf{v}^*; \alpha, \beta)). \tag{4.6}$$

The transition probability density models the probability density that a candidate particle in $\mathbf{x}$ modifies the velocity into that of the test particle due to the interaction with a field particle in the visibility domain. Its formal expression reads:

$$\mathcal{A}[f](\mathbf{v}_* \rightarrow \mathbf{v}|\mathbf{x}, \mathbf{x}^*, \mathbf{v}_*, \mathbf{v}^*; \alpha, \beta). \tag{4.7}$$

Step 3. Selection of the mathematical structure: Kinetic models are stated in terms of an evolution equation for the distribution function f, deduced as a balance law in the space of the microscopic states. Calculations analogous to those presented in Section 3.4 lead to the following structure:

$$\left(\frac{\partial}{\partial t} + \mathbf{v} \cdot \nabla_{\mathbf{x}}\right) f(t, \mathbf{x}, \mathbf{v}) = J[f](t, \mathbf{x}, \mathbf{v}; \alpha, \beta)$$

$$= \int_{D_{\mathbf{v}} \times D_{\mathbf{v}}} \int_{\Omega(\mathbf{x}, \mathbf{v}_*)} \eta[f](\mathbf{x}, \mathbf{x}^*, \mathbf{v}_*, \mathbf{v}^*; \alpha, \beta) \, \mathcal{A}[f](\mathbf{v}_* \to \mathbf{v}|\mathbf{x}, \mathbf{x}^*, \mathbf{v}_*, \mathbf{v}^*; \alpha, \beta)$$

$$\times f(t, \mathbf{x}, \mathbf{v}_*) f(t, \mathbf{x}^*, \mathbf{v}^*) \, d\mathbf{x}^* \, d\mathbf{v}^* \, d\mathbf{v}_*$$

$$- f(t, \mathbf{x}, \mathbf{v}) \int_{D_{\mathbf{v}}} \int_{\Omega(\mathbf{x}, \mathbf{v})} \eta[f](\mathbf{x}, \mathbf{x}^*, \mathbf{v}, \mathbf{v}^*; \alpha, \beta) \, f(t, \mathbf{x}^*, \mathbf{v}^*) \, d\mathbf{x}^* \, d\mathbf{v}^*, \quad (4.8)$$

where $\nabla_{\mathbf{x}}$ denotes the gradient operator with respect to the space variables, while operators modeling interactions express the *gain* and the *loss* of pedestrians in the elementary volume of the phase space around the test microscopic state $(\mathbf{x}, \mathbf{v})$. This structure is consistent with the assumptions presented in Section 3.4, namely that interactions involve three types of a-particles: *test particle*, *field particle*, and *candidate particle*, whose distribution functions are $f_i(t, \mathbf{x}, \mathbf{v})$, $f_k(t, \mathbf{x}, \mathbf{v}^*)$, and $f_h(t, \mathbf{x}, \mathbf{v}_*)$, respectively.

Let us now consider how specific models can be derived consistently with the structure defined by Eq. (4.8). The main reference for the reasoning proposed in the following is the recent paper which mainly focuses on multiscale methods [6]. Two further assumptions are made:

1. the encounter rate is constant, i.e., $\eta = \eta_0$; and

2. the transition probability density does not depend on the position and velocity of the field particle, but simply on quantities averaged over $\Omega(\mathbf{x}, \mathbf{v})$, i.e.,

$$\mathcal{A}[f](\mathbf{v}_* \to \mathbf{v}|\mathbf{x}, \mathbf{x}^*, \mathbf{v}_*, \mathbf{v}^*; \alpha, \beta).$$

Using these assumption, the structure (4.8) simplifies to:

$$\left(\frac{\partial}{\partial t} + \mathbf{v} \cdot \nabla_{\mathbf{x}}\right) f(t, \mathbf{x}, \mathbf{v}) = J[f](t, \mathbf{x}, \mathbf{v}; \alpha, \beta)$$

$$= \eta_0 \int_{D_{\mathbf{v}}} \mathcal{A}[f](\mathbf{v}_* \to \mathbf{v}|\mathbf{x}, \mathbf{v}_*; \alpha, \beta) f(t, \mathbf{x}, \mathbf{v}_*) \rho_{\mathbf{v}_*}(t, \mathbf{x}, \mathbf{v}_*) \, d\mathbf{v}_*$$

$$- \eta_0 \, \rho_{\mathbf{v}}(t, \mathbf{x}, \mathbf{v}) f(t, \mathbf{x}, \mathbf{v}), \quad (4.9)$$

where $\rho_{\mathbf{v}_*}(t, \mathbf{x}, \mathbf{v}_*)$ and $\rho_{\mathbf{v}}(t, \mathbf{x}, \mathbf{v})$ are the densities in the visibility domain related to the velocity directions $\mathbf{v}_*$ and $\mathbf{v}_*$, respectively:

$$\rho_{\mathbf{v}_*} = \rho_{\mathbf{v}_*}[f](t, \mathbf{x}, \mathbf{v}_*) = \int_{D_{\mathbf{v}}} \int_{\Omega(\mathbf{x}, \mathbf{v}_*)} f(t, \mathbf{x}^*, \mathbf{v}^*) d\mathbf{x}^* d\mathbf{v}^*, \quad (4.10)$$

and

$$\rho_{\boldsymbol{v}} = \rho_{\boldsymbol{v}}[f](t, \mathbf{x}, \boldsymbol{v}) = \int_{D_v} \int_{\Omega(\mathbf{x}, \boldsymbol{v})} f(t, \mathbf{x}^*, \mathbf{v}^*) d\mathbf{x}^* d\mathbf{v}^*. \tag{4.11}$$

As for the modeling of the velocity dynamics, interactions are supposed to trigger a decision process which comprises the following steps: selection of the walking direction and then adjustment of the walking speed; see [12] for further details. These steps are supposed to be sequential and dependent on the local flow conditions. A simple model consistent with these assumption is given by a product of delta distributions as follows:

$$\mathcal{A}[\rho, \boldsymbol{\xi}](\mathbf{v}_* \to \mathbf{v} | \mathbf{x}, \mathbf{v}_*; \alpha, \beta) = \mathcal{A}^v[\rho](v_* \to v | \mathbf{x}, \mathbf{v}_*; \alpha) \, \mathcal{A}^\theta[\rho, \boldsymbol{\xi}](\theta_* \to \theta | \mathbf{x}, \mathbf{v}_*; \beta)$$
$$= \delta\big(v - \varphi_*[\rho](\mathbf{x}, \mathbf{v}_*, \boldsymbol{\omega}_*; \alpha)\big) \, \delta\big(\theta - \omega_*[\rho, \boldsymbol{\xi}](\mathbf{x}, \mathbf{v}_*; \beta)\big), \tag{4.12}$$

where, following [6], $\boldsymbol{\omega}_*$ and φ_* are the preferred walking direction and speed, respectively.

Let us discuss in some more detail how one may model the decision process by which pedestrians select their preferred walking speed. As mentioned, it is expected that at high density walkers move away from overcrowded areas by choosing the direction of the direction of less congestion), $\boldsymbol{v}^{(v)}$, while at low density, walkers head for the target identified by $\boldsymbol{v}^{(t)}$, unless their level of stress is high in which case they tend to follow the mean pedestrian stream given by $\boldsymbol{v}^{(s)}$. Accordingly:

$$\boldsymbol{\omega}_* = \boldsymbol{\omega}_*[\rho, \boldsymbol{\xi}](\mathbf{x}, \mathbf{v}_*; \beta) = \frac{\rho_{\boldsymbol{v}_*} \boldsymbol{v}_*^{(v)} + (1 - \rho_{\boldsymbol{v}_*}) \dfrac{\beta \, \boldsymbol{v}_*^{(s)} + (1-\beta) \boldsymbol{v}_*^{(t)}}{\|\beta \, \boldsymbol{v}_*^{(s)} + (1-\beta) \boldsymbol{v}_*^{(t)}\|}}{\left\| \rho_{\boldsymbol{v}_*} \boldsymbol{v}_*^{(v)} + (1 - \rho_{\boldsymbol{v}_*}) \dfrac{\beta \, \boldsymbol{v}_*^{(s)} + (1-\beta) \boldsymbol{v}_*^{(t)}}{\|\beta \, \boldsymbol{v}_*^{(s)} + (1-\beta) \boldsymbol{v}_*^{(t)}\|} \right\|}. \tag{4.13}$$

Remark 4.2   This specific model has been derived based on the following assumptions on the interactions at the micro-scale.

1. The relative importance between $\boldsymbol{v}^{(s)}$ and $\boldsymbol{v}^{(t)}$ is determined by the parameter β, i.e., the weight of $\boldsymbol{v}^{(s)}$ increases as β increases.

2. The relative importance between $\boldsymbol{v}^{(v)}$ and the direction defined in Item 1 is determined by the perceived density $\rho_{\boldsymbol{v}}$, i.e., the weight of $\boldsymbol{v}^{(v)}$ increases as $\rho_{\boldsymbol{v}}$ increases.

3. The change of speed depends on the difference between the perceived densities after and before the change of the velocity direction. This dynamics is enhanced by the parameter α.

Let us now briefly discuss how boundary conditions can be formulated. Different modeling approaches can be used; see [14] and [6]. The general idea is that pedestrians modify their walking trajectory so as to avoid the contact with walls. More specifically, one may assume that the preferred walking direction θ is chosen in two steps. First, a direction $\boldsymbol{\omega}_1$, identified by the angle θ_1, is selected by weighting the directions $\boldsymbol{\nu}^{(v)}$, $\boldsymbol{\nu}^{(t)}$, and $\boldsymbol{\nu}^{(s)}$, as already described. If this walking direction points toward an exit area, then it is not modified, namely $\boldsymbol{\omega}_1 = \boldsymbol{\omega}$. Otherwise, the preferred walking direction is modified by a weighted choice between θ_1 and the direction $\theta^{(t)}$ of $\boldsymbol{\nu}^{(t)}$, where the weight is given by the distance from the closest wall. Accordingly, the transition probability density for the walking direction is defined as follows:

$$\mathcal{A}^{\theta}[\rho, \boldsymbol{\xi}](\theta_* \rightarrow \theta | \mathbf{x}, \mathbf{v}_*; \beta) = \delta\left(\theta - \theta_*\right), \quad \text{with} \quad \theta = (1 - d_w)\theta^{(t)} + d_w \theta_1, \qquad (4.14)$$

where θ_1 can be computed as in Eq. (4.13) and d_w is taken to be equal to one if θ_1 is directed toward an exit area and to the distance from the closest wall otherwise.

Note that, although the walking strategy encompasses the tendency to keep distance from walls, some pedestrians, in probability, might reach the boundaries and, therefore, an appropriate scattering model needs to be given. As an example, in [6] it is assumed that at the walls pedestrians keep their speed constant but modify the velocity direction, making it tangent to the wall. By imposing zero-net flux at the wall, the following condition can be thus obtained:

$$f(t, \mathbf{x}_B, \mathbf{v})|\mathbf{v} \cdot \mathbf{n}_B| = \int_{\mathbf{v}_* \cdot \mathbf{n}} \delta\left(v - \varphi_*[\rho](\mathbf{x}_B, \mathbf{v}_*, \boldsymbol{\nu}_B^{(t)}; \alpha)\right) \delta\left(\boldsymbol{\nu} - \boldsymbol{\nu}_B^{(t)}\right) f(t, \mathbf{x}_B, \mathbf{v}_*)|\mathbf{v}_* \cdot \mathbf{n}_B| dv_*,$$

$$(4.15)$$

where $\mathbf{n}_B$ and $\boldsymbol{\nu}_B^{(t)}$ are the unit vectors orthogonal to the wall and directed toward the target at $\mathbf{x}_B \in \partial\Sigma$.

4.5 SAMPLE SIMULATIONS

As discussed in Section 1.3, the kinetic modeling approach leads to models which are mathematically involved. Therefore, the numerical simulation is the only viable mean to assess their predictive ability. The sample simulations presented in the following mainly focus on how the level of stress affects the crowd dynamics. As already stressed, it is assumed that this level of stress is shared by all the pedestrians in the crowd and stay constant in time. The modeling of the onset and propagation of stressful conditions is treated in the next chapter.

Two case studies have been specifically selected out of the many situations which may normally occur in urban environments. First, the interaction of two groups of people moving in opposite directions along a narrow street. Second, the evacuation dynamics of commuters who are waiting for a train along the platform of a railway station. These situations are shown in Figs. 4.1 and 4.2.

The simulations described below were carried out by a Monte Carlo particle method and refer to a kinetic model which presents minor technical variations with respect to the one discussed in the preceding section. More specifically, interactions are treated as local in space, i.e., macroscopic quantities in the visibility domain simplify to their local values. Furthermore,

Figure 4.1: Counterflow in an overcrowded city street.

pedestrians are supposed to select the preferred walking direction according to Eq. (4.13) and the preferred walking speed by a decision process which was designed to lead to a collective emerging behavior consistent with the velocity-speed diagram for homogeneous crowd in steady conditions as observed experimentally. The interested reader is referred to [30] for a more detailed description of the model.

The first sample simulation refers to the dynamics of two groups of people moving along opposite directions along a narrow street. The main aim is to assess the capability of the model to depict the so-called *segregation dynamics in counter-flows* which is an emerging behavior commonly observed in the pedestrian dynamics. General aspects of self-organization have been studied by various authors, for instance [30–32, 36], as it is considered an interesting benchmark to validate models. In more detail, references [31, 32] aim at showing that self-organization results from avoiding collisions while a more general framework is proposed in [30], where it has been shown how self-organization is the output of a complex ensemble of interactions.

Initially, the two groups of pedestrians are uniformly randomly distributed over the street and move toward its opposite sides. Boundary conditions are assumed to be periodic, i.e., the simulation domain is a small portion of the street.

Simulations show that pedestrians segregate gradually in time into two main lanes of uniform walking direction, namely the group moving rightward gather in the upper part of the street while the group moving leftward gather in the lower part. This is clearly shown in the panels (a) and (b) of Fig. 4.3. Note that this is an emerging behavior which is not triggered by an

Figure 4.2: Commuters on an underground train platform.

ad-hoc modeling of interactions, namely pedestrians moving in opposite way interact according the same rules of pedestrians moving in the same direction. Interestingly, panels (c) and (d) of Fig. 4.3 shows that, under stressful conditions, this self-organized behavior doesn't emerge any more and pedestrians randomly fill the street.

The role played by the level of stress in the dynamics of pedestrians can be more precisely quantified by computing the so-called *band index* [12]:

$$Y_{\mathrm{B}}(t) = \frac{1}{l_x l_y} \int_0^{l_y} \left| \int_0^{l_x} \frac{\rho_A(t, \mathbf{x}) - \rho_B(t, \mathbf{x})}{\rho_A(t, \mathbf{x}) + \rho_B(t, \mathbf{x})} dx \right| dy, \tag{4.16}$$

where l_x, l_y are the longitudinal and transversal lengths of the street, and ρ_1, ρ_2 are the density of the two groups of pedestrians. According to Eq. (4.16), $Y_{\mathrm{B}}(t) = 0$ for mixed counter flows while $Y_{\mathrm{B}}(t) = 1$ for perfect segregation.

Figure 4.4 shows that the band index reaches an asymptotic value after a transient time for every value of the level of stress, β. However, it is apparent that the segregation of pedestrians is anticipated to be more pronounced for $\beta = 0.5$ while it drops for higher level of stress. Therefore, there exists a value of the level of stress which lead pedestrian to behave in the most rational way, namely $\beta = 0.5$. Indeed, for lower values of β pedestrians are not motivated to follow the trajectory which permits to get the target in the most efficient way, while for larger values of β the irrational behavior prevail on the rational one and, therefore, also in this case, pedestrians do not follow the optimal trajectory.

The second sample simulation refers to the evacuation dynamics of a crowd, namely the case of commuters who have to leave the platform of a railway station.

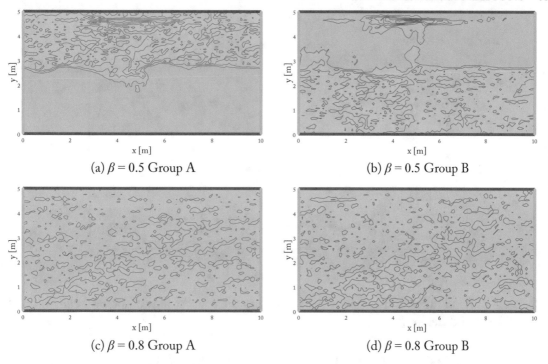

(a) $\beta = 0.5$ Group A

(b) $\beta = 0.5$ Group B

(c) $\beta = 0.8$ Group A

(d) $\beta = 0.8$ Group B

Figure 4.3: Density contour plot of a crowd composed of 50 pedestrians divided into 2 groups which move leftward and rightward.

The main aim is to assess the capability of the model to provide a reasonable description of pedestrians behavior in a situation of emergency. This latter is a crucial requirement for crowd dynamics models which are expected to be the core of simulation platforms designed to support urban planners and/or crisis managers. Indeed, these platforms would be of paramount importance in that they may offer a virtual and augmented-reality environment which permit one to improve the management of safety problems and/or the design of buildings [13], as well as to train crisis managers by allowing them to explore different scenarios resulting from possible actions. Some perspective ideas are proposed in [11] toward the development of machine learning devices to select optimal safety actions by means of analyzing a database repository of a large number of simulations corresponding to different control actions.

Initially, individuals are waiting at rest by occupying uniformly the platform as shown in panel (a) of Fig. 4.5. The crowd dynamics leads to the onset of overcrowded regions close to the areas where there is an abrupt change of geometry and these region grow in time. This is apparent in panel (b) of Fig. 4.5. The overcrowding occurs because pedestrians are close to the exit and this weakens their tendency of avoiding the more congested areas.

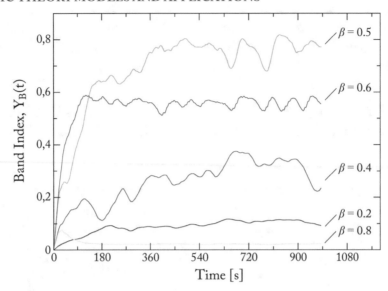

Figure 4.4: Band index of the density field related to the counter flow of a crowd composed of 50 walkers for different β.

4.6   A BRIDGE TOWARD RESEARCH PERSPECTIVES

The contents of this chapter have shown that there is a growing interest in the kinetic theory approach to the modeling of human crowds. The rapid development of kinetic models has been fueled by encouraging results [30] as well as by the potential capability of these models to provide a realistic description of crowd dynamics.

In spite of the progress made, many difficult challenges lie ahead for crowd modeling. Indeed, there are still many key requirements which need to be properly fulfilled. As pointed out by crisis managers, computational models should permit one to carry out simulations in complex environments constituted of several interconnected areas, each of them presenting different geometrical features. Furthermore, models should allow for the possibility of assessing the consequences of external actions, either vocal or visual, which can be used to steer pedestrians during evacuation. Safety problems also motivate the study of the onset and propagation of stress conditions in crowds as well as how this spreading may affect the overall crowd behavior. Recent achievements on behavioral crowds [14, 53] have provided significant contributions to this topic but many others still need to come.

Two additional very important research topics can be briefly mentioned. First, a more fundamental understanding of interactions between pedestrians is needed, as well as their possible dependence on the local flow conditions. As an example, it is likely that the higher is the local density, the higher becomes the level of stress. Second, the multiscale modeling approach to pedestrians crowds is needed because no single observation and representation scale can fully

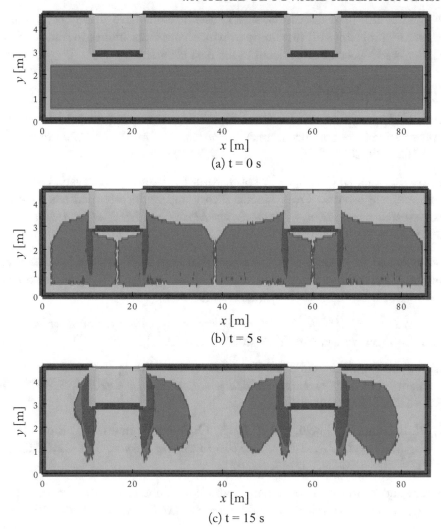

(a) t = 0 s

(b) t = 5 s

(c) t = 15 s

Figure 4.5: Density contour plot of a crowd composed of 200 pedestrians evacuating the platform of a metro station at different times.

capture the collective dynamics of living systems. Indeed, the dynamics at the microscopic scale paves the way to the derivation of models at the mesoscopic scale. In turn, the hydrodynamic models, which are based on observable macroscopic quantities, can be obtained from kinetic models by letting the distance between individuals tend to zero.

To conclude this section, it is worth stressing that the research activity in the field is rapidly growing due to the potential impact that can have on dealing with safety problems.

As also highlighted by this brief discussion, several useful results have already been achieved, but many challenging problems still remain open which constitute interesting lines of future research. The latter is addressed in more detail in the next chapter.

4.7   BIBLIOGRAPHY

[1] J. P. Agnelli, F. Colasuonno, and D. Knopoff, A kinetic theory approach to the dynamics of crowd evacuation from bounded domains, *Mathematical Models and Methods in Applied Sciences*, 25:109–129, 2015. DOI: 10.1142/s0218202515500049 52

[2] G. Albi, N. Bellomo, L. Fermo, S.-Y. Ha, J. Kim, L. Pareschi, D. Poyato, and J. Soler, Traffic, crowds, and swarms. From kinetic theory and multiscale methods to applications and research perspectives, *Mathematical Models and Methods in Applied Sciences*, 29(10):1901–2005, 2019. DOI: 10.1142/S0218202519500374 52, 54

[3] A. Alekseenko and E. Josyula, Deterministic solution of the Boltzmann equation using discontinuous Galerkin discretization in velocity space, *Journal of Computational Physics*, 272:170–188, 2014. DOI: 10.1016/j.jcp.2014.03.031 54

[4] V. V. Aristov, *Direct Methods for Solving the Boltzmann Equation and Study of Nonequilibrium Flows*, Springer-Verlag, New York, 2001. DOI: 10.1007/978-94-010-0866-2 54

[5] V. V. Aristov and F. G. Tcheremissine, The conservative splitting method for the solution of a Boltzmann, *U.S.S.R Computational Mathematical Physics*, 20:208–225, 1980. DOI: 10.1016/0041-5553(80)90074-9 54

[6] B. Aylaj, N. Bellomo, L. Gibelli, and A. Reali, On a unified multiscale vision of behavioral crowds, *Mathematical Models and Methods in Applied Sciences*, 30:1–22, 2020. DOI: 10.1142/S0218202520500013 55, 58, 59, 60

[7] H. Babovsky and R. Illner, A convergence proof for Naubu's simulation method for the full Boltzmann equation, *Mathematical Methods of Applied Sciences*, 8:223–233, 1986. DOI: 10.1137/0726004 55

[8] N. Bellomo and A. Bellouquid, On multiscale models of pedestrian crowds from mesoscopic to macroscopic, *Communications in Mathematical Sciences*, 13(7):1649–1664, 2015. DOI: 10.4310/cms.2015.v13.n7.a1 53

[9] N. Bellomo, A. Bellouquid, L. Gibelli, and N. Outada, *A Quest Towards a Mathematical Theory of Living Systems*, Birkhäuser-Springer, New York, 2017. DOI: 10.1007/978-3-319-57436-3

[10] N. Bellomo, A. Bellouquid, and D. Knopoff, From the micro-scale to collective crowd dynamics, *Multiscale Modelling and Simulation*, 11:943–963, 2013. DOI: 10.1137/130904569 52, 53

[11] N. Bellomo, D. Clarke, L. Gibelli, P. Townsend, and B. J. Vreugdenhil, Human behaviours in evacuation crowd dynamics: From modelling to "big data" toward crisis management, *Physics of Life Reviews*, 18:1–21, 2016. DOI: 10.1016/j.plrev.2016.05.014 63

[12] N. Bellomo and L. Gibelli, Toward a behavioral-social dynamics of pedestrian
crowds, *Mathematical Models and Methods in Applied Sciences*, 25:2417–2437, 2015. DOI:
10.1142/S0218202515400138 53, 59, 62

[13] L. Gibelli and N. Bellomo, Eds., *Crowd Dynamics, Volume 1: Theory, Models, and Safety
Problems*, Birkhäuser-Springer, New York, 2017. DOI: 10.1007/978-3-030-05129-7 63

[14] N. Bellomo, L. Gibelli, and N. Outada, On the interplay between behavioral dynamics
and social interactions in human crowds, *Kinetic and Related Models*, 12:397–409, 2019.
DOI: 10.3934/krm.2019017 53, 55, 59, 64

[15] A. L. Bertozzi, J. Rosado, M. B. Short, and L. Wang, Contagion shocks in one dimension,
Journal Statistical Physics, 158(3):647–664, 2015. DOI: 10.1007/s10955-014-1019-6 53

[16] G. A. Bird, *Molecular Gas Dynamics and the Direct Simulation of Gas Flows*, Oxford University Press, 1994. 54

[17] A. Corbetta, L. Bruno, A. Mountean, and F. Yoschi, High statistics measurements of
pedestrian dynamics, models via probabilistic method, *Transport Research Proceedings*,
2:96–104, 2014. DOI: 10.1016/j.trpro.2014.09.013 53

[18] A. Corbetta, A. Mountean, and K. Vafayi, Parameter estimation of social forces in pedestrian dynamics models via probabilistic method, *Mathematical Biosciences Engineering*,
12:337–356, 2015. DOI: 10.3934/mbe.2015.12.337 53

[19] P. Degond, C. Appert-Rolland, M. Moussaid, J. Pettré, and G. Theraulaz, A hierarchy
of heuristic-based models of crowd dynamics, *Journal Statistical Physics*, 152:1033–1068,
2013. DOI: 10.1007/s10955-013-0805-x 53

[20] P. Degond, C. Appert-Rolland, J. Pettré, and G. Theraulaz, Vision based macroscopic pedestrian models, *Kinetic and Related Models*, 6(4):809–839, 2013. DOI:
10.3934/krm.2013.6.809 53

[21] G. Dimarco and L. Pareschi, Numerical methods for kinetic equations, *Acta Numerica*,
23:369–520, 2014. DOI: '10.1017/s0962492914000063 54

[22] A. Elmoussaoui, P. Argoul, M. ElRhabi, and A. Hakim, Discrete kinetic theory for
2-D modeling of a moving crowd: Application to the evacuation of a non-connected
bounded domain, *Computers Mathematics with Applications*, 75:1159–1180, 2018. DOI:
10.1016/j.camwa.2017.10.023 53

[23] S. Faure and B. Maury, Crowd motion from the granular standpoint, *Mathematical Models and Methods in Applied Sciences*, 25:463–493, 2015. DOI: 10.1142/s0218202515400035 53

[24] L. Fermo and A. Tosin, Fundamental diagrams for kinetic equations of traffic flow, *Dynamical Systems Series S*, 7(3):449–462, 2014. DOI: 10.3934/dcdss.2014.7.449 53

[25] J. H. Ferziger and M. Peric, *Computational Methods for Fluid Dynamics*, Springer Science and Business Media, 2012. DOI: 10.1007/978-3-642-56026-2 54

[26] F. Filbet, C. Mouhot, and L. Pareschi, Solving the Boltzmann equation in $N \log 2N$, *SIAM Journal of Scientific Computing*, 28:1029–1053, 2006. DOI: 10.1137/050625175 54

[27] A. Frezzotti, Numerical study of the strong evaluation of a binary mixture, *Fluid Dynamics Research*, 8:175–187, 1991. DOI: 10.1016/0169-5983(91)90041-G 54

[28] I. M. Gamba and S. H. Tharkabhushanam, Spectral-Lagrangian methods for collisional models of non-equilibrium statistical states, *Journal of Computational Physics*, 228:2012–2036, 2009. DOI: 10.1016/j.jcp.2008.09.033 54

[29] G. P. Ghiroldi and L. Gibelli, A direct method for the Boltzmann equation based on a pseudo-spectral velocity space discretization, *Journal of Computational Physics*, 258:568–584, 2014. DOI: 10.1016/j.jcp.2013.10.055 54

[30] N. Bellomo and L. Gibelli, Behavioral crowds: Modeling and Monte Carlo simulations toward validation, *Computers and Fluids*, 141:13–21, 2016. DOI: 10.1016/j.compfluid.2016.04.022 53, 55, 61, 64

[31] G. H. Goldsztein, Moving around a two-lane circular track in both directions. Avoiding collisions leads to self-organization, *SIAM Journal Applied Mathematics*, 76:1433–1445, 2016. DOI: 10.1137/140996732 61

[32] G. H. Goldsztein, Self-organization when pedestrians move in opposite directions. Multi-lane circular track model, *Applied Sciences*, 10(563), 2020. DOI: 10.3390/app10020563 61

[33] M. S. Ivan and S. V. Rogazinsky, Theoretical analysis of traditional and modern schemes of the DSMC method, *Proc. of the 9th International Symposium on Rarefied Gas Dynamics*, A. E. Beylich, Ed., 2:629–642, 1990. 55

[34] D. Kim and A. Quaini, A kinetic theory approach to model pedestrian dynamics in bounded domains with obstacles, *Kinetic and Related Models*, 12(6):1273–1296, 2019. DOI: 10.3934/krm.2019049 53

[35] K. Koura, Null-collision technique in the direct simulation Monte Carlo technique, *Physics of Fluids*, 29:3509–3511, 1986. DOI: 10.1063/1.865826 55

[36] T. Kretz, A. Grünebohm, M. Kaufman, F. Mazur, and M. Schreckenberg, Experimental study of pedestrian counterflow in a corridor, *Journal Statistical Mechanics Theory Experiments*, 1001, 2006. DOI: 10.1088/1742-5468/2006/10/p10001 61

[37] B. Maury and J. Venel, A discrete contact model for crowd motion, *ESAIM: Mathematical Models Numerical Analysis*, 45:145–168, 2011. DOI: 10.1051/m2an/2010035 53

[38] C. Mouhot and L. Pareschi, Fast algorithms for computing the Boltzmann collision operator, *Comptes Rendus de l'Académy de Sciences Paris*, 339:71–76, 2004. DOI: 10.1090/s0025-5718-06-01874-6 54

[39] M. Moussaïd, D. Helbing, S. Garnier, A. Johansson, M. Combe, and G. Theraulaz, Experimental study of the behavioural mechanisms underlying self-organization in human crowds, *Proc. Royal Society B*, 276:2755–2762, 2009. DOI: 10.1098/rspb.2009.0405 53

[40] K. Nanbu, Direct simulation scheme derived from the Boltzmann equation, I. Monocomponent gases, *Japan Journal of Physics*, 19:2042–2049, 1980. DOI: 10.1143/JPSJ.49.2042 55

[41] A. Nordsieck and B. Hicks, Monte Carlo evaluation of the Boltzmann collision integral, *Proc. of the 9th International Symposium on Rarefied Gas Dynamics*, C. L. Brudin, Ed., 2:695–710, 1967. 54

[42] T. Ohwada, Structure of normal shock wave: Direct numerical analysis of the Boltzmann equation for hard-sphere molecules, *Physics of Fluids*, 5(1):217–234, 1992. DOI: 10.1063/1.858777 54

[43] L. Pareschi and G. Russo, Numerical solution of the Boltzmann equation. I. Spectrally accurate approximation of the collision operator, *SIAM Journal of Numerical Analysis*, 37(4):1217–1245, 2000. DOI: 10.1137/s0036142998343300 54

[44] L. Pareschi, and G. Toscani, *Interacting Multiagent Systems: Kinetic Equations and Monte Carlo Methods*, Oxford University Press, Oxford, 2013. 54

[45] S. Rjasanow and W. Wagner, *Stochastic Numerics for the Boltzmann Equation*, Springer, Berlin, 2005. DOI: 10.1007/3-540-27689-0 55

[46] A. Schadschneider, M. Chraibi, A. Seyfried, A. Tordeux, and J. Zhang, Pedestrian dynamics: From empirical results to modeling, in *Crowd Dynamics Volume 1: Theory Models and Safety Problems*, L. Gibelli and N. Bellomo, Eds., pages 63–102, Birkhäuser, Springer Nature, 2018. DOI: 10.1007/978-3-030-05129-7_4 53

[47] A. Schadschneider and A. Seyfried, Empirical results for pedestrian dynamics and their implications for cellular automata models, *Pedestrian Behavior-Models, Data Collection, and*

Applications, H. Timmermans, Ed., Chapter 2, pages 27–44, Emerald Group Publishing, 2009. DOI: 10.1108/9781848557512-002 52

[48]   A. Schadschneider and A. Seyfried,   Empirical results for pedestrian dynamics and their implications for modeling, *Networks Heterogenous Media*, 6:545–560, 2011. DOI: 10.3934/nhm.2011.6.545 52, 53

[49]   A. Seyfried, B. Steffen, W. Klingsch, and M. Boltes, The fundamental diagram of pedestrian movement revisited, *Journal Statistical Mechanics: Theory and Experiments*, 360:232–238, 2006. DOI: 10.1088/1742-5468/2005/10/p10002 52, 53

[50]   Y. Sone, T. Ohwada, and K. Aoki, Temperature jump and Knudsen layer in a rarefied gas over a plane wall: Numerical analysis of the linearized Boltzmann equation for hard-sphere molecules, *Physics of Fluids*, 1:363–370, 1989. DOI: 10.1063/1.857457 54

[51]   S. K. Stefanov,   On DSMC calculations of rarefied gas flows with small number of particles in cells,   *SIAM Journal of Scientific Computing*, 33:677–702, 2011. DOI: 10.1137/090751864 55

[52]   W. Wagner,   A convergence proof of Bird's direct simulation Monte-Carlo method for the Boltzmann equation,   *Journal of Statistical Physics*, 66:1011–1044, 1992. DOI: 10.1007/BF01055714 54

[53]   L. Wang, M. Short, and A. L. Bertozzi, Efficient numerical methods for multiscale crowd dynamics with emotional contagion, *Mathematical Models and Methods in Applied Sciences*, 27:205–230, 2017. DOI: 10.1142/s0218202517400073 51, 64

[54]   V. Yanitskiy, Operator approach to direct simulation Monte Carlo theory in rarefied gas dynamics, *Proc. of the 9th International Symposium on Rarefied Gas Dynamics*, A. E. Beylich, Ed., 2:770–777, 1990. 55

CHAPTER 5

Kinetic Theory Models Toward Research Perspectives

5.1   INTRODUCTION

Abstract: This last chapter is devoted to go through the topics presented in the book, with the aim of looking forward to research perspectives. Key ideas such as the multiscale vision of crowd dynamics, pattern formation, and management of crisis events, are deepened. The chapter, and the book, concludes with the formulation of three relevant questions that—according to the authors' opinion—the research future activity will strive to address.

A number of possible research perspectives have already been introduced in the preceding chapters, generally in the last section of each chapter. This present chapter revisits the various hints to provide an overall presentation toward a research perspectives which look ahead to a unified quest toward research programs. The selection of the topics does not claim to be exhaustive as it is guided by the authors' bias and research experience.

In detail, the first part of the chapter focuses on the following three *key topics*:

1. a multiscale vision of crowd dynamics, where all scales contribute to the modeling approach;

2. social interactions which modify the activity variable and propagate emotional states in space; and

3. a systems approach to behavioral crowds followed by some reasonings on safety problems.

These topics are critically analyzed within a qualitative framework which avoids technical formalizations which can be effectively achieved only within well-specialized research projects.

The second part provides some free reasonings to identify a number of selected research perspectives which will arguably attract the interest of applied mathematicians and physicists. Three *key questions* are selected, once more guided by the authors' bias, while the answer to them should "hopefully" promote research activity in the filed.

The presentation differs from that of the preceding chapters by the lack of mathematical formalization which is replaced by concepts and qualitative suggestions. The contents of the chapter is as follows. Section 5.2 presents some conceptual ideas on a multiscale vision of crowd dynamics first by showing how the same principles and analogous parameters can be used to derive crowds models at each scale and, subsequently, how models at the higher scale can be derived from the underlying description at the low scale. Section 5.3 proposes studying the behavioral features which might appear in crowds and subsequently on the influence over the motion and pattern formation. Section 5.4 focuses, looking ahead to safety problems, on the derivation of a systems approach to crowds dynamics accounting for motions across a network of areas each of them characterized by different geometrical and physical features. The achievements treated in Section 5.3 are supposed to provide the necessary conceptual framework for this specific objective. Section 5.5 proposes a closure where we take the liberty of selecting some research perspective which, according to the authors' bias, will pervade research activity on crowds modeling in the next decade.

5.2   ON A MULTISCALE VISION

Although this Lectures Note has been mainly focused on the kinetic theory approach, it has been mentioned, right from the beginning, that the modeling of crowds can be developed at each of the three conventional scales, namely microscopic (individual based), mesoscopic (kinetic), and macroscopic (hydrodynamical). The strategic selection of a specific modeling scale has been discussed in Chapter 2, where our motivations toward the kinetic theory approach have been reported within the framework of a general critical analysis of the literature in the field.

In addition to all reasonings developed in Chapter 2, we believe that the approach should go beyond the technical selection of one specific scale as it should look forward to a multiscale vision somehow inspired by real world applications. Some aspects of a multiscale vision have been treated in [4] which provide the conceptual background, supporting the contents of this section.

In more detail, the authors have shown in [4] that the modeling at each scale can be developed within a conceptual framework where the same behavioral features are selected toward the modeling approach at each scale. In addition, each feature is characterized by analogous parameters. This strategy is useful to develop a systems approach for crowds which move across venues constituted by different interconnected areas, where each area might require a different scale according to different geometrical and computational specificity. Hence, crossing different areas might also suggest moving between diferent representation-modeling scales. The common parametrization at each scale makes simpler the overall computational treatment.

In more detail, the approach refers to models where the velocity dynamics is described by a hierarchy in the decision. Namely, walkers first select the velocity directions and subsequently adapt the speed to the new density conditions. The selection of the velocity direction is the same, at each scale, corresponding to the modeling strategy used in the kinetic theory approach reported in Chapter 4. Namely, walkers select the said direction as a weighted choice from trend to the target, avoiding walls and overcrowded area, and attraction to the stream. The technical difference is that the strategy is applied to individuals at the micro-scale, to the statistical representative in the kinetic theory approach, and to individuals in the elementary volume of the physical space in the case of the macro-scale.

The weight in the selection is applied, as we have seen in Chapter 4, by the local density and by a parameter modeling the stress level. The adaptation of the speed to the new direction is referred to the new density conditions. Namely, walkers reduce the speed when they move to increasing conditions and increase the speed, when they move to lower-density conditions. Models refer, in addition to density conditions, also to the parameter modeling the quality.

Going beyond the purely phenomenological approach, the derivation of models can be achieved by a micro-macro derivation inspired by the sixth Hilbert problem [22]. A possible generalization has been developed in [14, 15], however limited to the derivation of diffusion models from a kinetic theory model of large systems of interacting living entities.

The methodological approach proposed in [15] mimic an Hilbert-type expansion in terms of a small parameter related to the mean distance between pedestrians. A key problem in the approach is the definition of an equilibrium state to develop the expansion. This state, in the case of classical particles is delivered by the Maxwellian distribution, while in the case of crowds can be given by steady uniform flows. This need motivates the search of analytic expression of the said equilibrium distribution as developed, for instance, in [9].

This type of technique has been developed in the case of vehicular traffic [6] and for crowd models in unbounded domains [5]. The problem in domains with walls and obstacles is still open. The final target would be using models at the micro-scale to implement the approach by tools of statistical physics, and derive hydrodynamical models from the aforementioned asymptotic analysis.

5.3   SOCIAL DYNAMICS AND PATTERN FORMATION IN CROWDS

Crowds should be viewed, according to some concepts introduced in Chapter 4, as a *behavioral systems*, where the rules by which walkers move depend on their social-emotional state which is heterogeneously distributed within the various populations composing the crowd. Hence, we used the term *behavioral dynamics* which has been made precise in the survey [25], where a detailed study of a broad variety of applications has been developed referring mainly to natural sciences, but also to applied sciences in general, see also [1].

The concept of *behavioral dynamics* has been made precise in [25] which presents an interesting review of a broad variety of applications developed by scientists, active in different branches of science. The authors also develop some statistics on the frequency of the use of methods suitable to account for behavioral features. Their study shows how the different approaches move toward a common research field which is progressively capturing the interest of the scientific community active on the study of real problems of interest in our society.

As we have seen in Chapter 5, the kinetic theory approach naturally accounts for the role of behavioral features which can modify substantially flow patterns. However, social behaviors cannot be confined to parameters, but they should be viewed as micro-scale variables which can be modified by interactions.

As an example to be properly developed, we can consider the approach proposed in [12] devoted to the modeling of propagation of stress caused by localized incidents. The modeling of interactions has been based on the heuristic assumption of the following sequence:

1. A walking strategy induced by a sequence of decisions where walkers first exchange their emotional state, then choose a walking direction, and finally adapt the speed to the new flow conditions.

2. The dynamics of the emotional state is modeled by theoretical tools of stochastic game theory by dynamics where individual carriers of a lower level of stress are attracted by individuals with a higher level of stress.

3. The level of stress is modeled by a variable with values in [0, 1] (the limits correspond, respectively, to the lowest and highest admissible levels). This variable depends on time and space by the consensus dynamics outlined in Item 2. Then, at each time, it acts locally in space with the same role of the stress introduced in Chapter 4.

Therefore, this approach differs from that presented in Chapter 4 only in the first step, where theoretical tools of game theory [2] are used to model the social dynamics. Simulations have shown how propagation of stress generates a motion which is far more disordered than that induced by a level of stress equally shared by all individuals. As we have seen, increasing stress drives the crowd toward irrational behaviors which can induce unsafe conditions.

However, the approach proposed in [12], see also [35], focuses only on a specific example, while its generalization is not straightforward. Some hints, selected among various possible ones, are given in the following as a contribution to a possible modeling strategy to be developed within a general framework.

• A selection of the specific social dynamics which can appear in crowds should be developed and guide the subdivision into functional subsystems corresponding to populations that express the specific dynamics. As an example, the crowd can be divided into leaders and followers, where leaders have been trained to behave rationally in critical situations, while followers are attracted to the rationality of leaders, to avoid irrational behaviors in the interactions among themselves.

• Modeling interactions for each specific social dynamic should be developed going beyond the consensus dynamics. As an example, the modeling of interactions proposed in [3] suggest taking into account both consent and dissent dynamics, while both dynamics are weighted by a parameter. Thus, developing theoretical tools of game theory [16, 23, 30] account for this important aspect of living systems interactions.

- Analogous social interactions can be modeled to account how external actions, for instance signaling, vocal, and visual communications, can interact with crowds with the aim of obtaining a consensus toward rational behaviors.

These concise indications show that the topic is open to development which can contribute to safety problems.

5.4   ON A SYSTEMS APPROACH TOWARD CRISIS MANAGEMENT

An important motivation for studying crowd dynamics already has been indicated in Chapter 1 focusing on the perspective that modeling and simulations can contribute to safety in situations generated by crisis situations whenever the dynamic generates overcrowding and loss of rationality in selecting the walking trajectory. In these situations, crisis managers are asked to select the most appropriate strategy to reduce danger. The contents of this section takes advantage of some preliminary ideas presented in [7] and [10] which have been further developed in [11].

The first step to tackling this type of problem is the design of a systems approach to crowd modeling to account for the complex features of crowds, with special focus on irrational behaviors, in complex venues. Subsequently, this approach can be addressed to support the needs of decision makers during a crisis situation.

Bearing all the above targets in mind, let us indicate three specific features, of the systems approach, selected among several possible ones. For each of them, simple examples are reported in italics.

1. Different qualities of walkers must be included in the modeling approach. *For instance, individuals with limited mobility, individuals with advanced mobility, and leaders trained to attract other individuals to rational behaviors.*

2. Dynamics of the crowds across venues constituted by interconnected areas, each of them with different geometrical and quality properties. *For example, the geometry can differ from long, narrow corridors to large rooms with or without internal obstacles; it can include going up and down stairs, different qualities of lighting, presence of smoke, and various other features [17, 33].*

3. The geometry of the system can be technically modified to account for modifications induced by incidents. *For instance, the closure of one of the exits due to a fire.*

The mathematical tools reported in Chapter 4 can, in principle, lead to a systems approach suitable to account for the requirements posed in the above items. In fact, the subdivision into functional subsystems can account for the different typologies of individuals in a crowd, while the parameters α and β can be toned for each component of the venue and labeled for each of them. These parameters can be labeled by a well-defined value corresponding to each component.

Moving across venues implies moving through different values of the said parameter. The passage from one value to the other can be sharp or smooth with respect to the space variable depending on the specific feature of the venue.

The approach should, in general, account for the use of empirical data toward modeling [34] and related computational problems [13] within a general framework of representation of real dynamics by virtual reality [24]. In addition, the problem of propagation of emotional state might be studied to account also for propagation in complex venues.

The third item appears to be far more complex, requiring not only advanced mathematical tools derived within the aforementioned systems approach, but also development of artificial intelligence tools related to computational modeling. Indeed, it is a challenging research perspective, where tools from mathematical and computer sciences can be addressed to support crisis managers and designers of venues according to safety requirements as enlightened in a rapidly growing bibliography, for example [7, 27, 31, 32, 37].

Before tackling this matter, let us define some possible objectives of a strategy toward artificial intelligence for crisis managers. In more detail, the computational model, once properly tuned, can be used toward the aims selected, among various possible ones, defined below, where some remarks have been added for each of them.

- **Design of safe global architectures:** Simulations can test different designs of the overall architecture of buildings venues with the aim of investigating how improved design can improve safety in critical conditions. Focusing on evacuation dynamics, the term "safety" is used here to identify local density below a threshold and evacuation time as much reduced as possible consistently with the former requirement.

- **Training crisis managers:** Can be planned by a design of a database where a big number of simulations is stored corresponding to different venues, crowd features, and actions to control crisis. Crisis managers can work out, for each specific venue, the most appropriate actions deemed to improve safety.

- **Selection of actions to support crisis situations:** The design of the database can be further developed to select, using artificial intelligence, the most appropriate actions to control a crisis by a detailed study of the flow patterns and of all aforementioned properties. The test might even study flow patterns induced by modifications of the quality and the geometry of the venue caused by specific incidents.

The various methods to treat these large amounts of data mentioned in the above items still need to be properly developed to define an emerging data science which aims at improving the decision-making process toward cost reductions and reduced risk. Therefore, the approach should go beyond the technical problem of data compression and their statistical interpretation.

5.5 CLOSURE: NEW TRENDS IN CROWD DYNAMICS

The five chapters of this Lecture Note have presented a constructive approach to the modeling and simulations of human crowds in complex venues. The approach presented here requires not only tools of physics and mathematics, but also additional knowledge concerning social and psychological behaviors. It has been mainly based on the mathematical tools of the kinetic theory for active particles within a multiscale vision, where the connection with the lower, individual-based, and higher, hydrodynamical scales is an important feature to be accounted for.

All chapters have introduced, generally in their last sections, the presentation of open problems to be properly developed within possible research programs. These perspectives include challenging analytic problems. These need new conceptual ideas, as well as new mathematical tools, to be properly tackled in the trend of mathematics for living systems.

This chapter has presented three key problems which were treated in the preceding sections where each of them has been followed by hints toward research perspectives. This final section can be viewed as a closure which, rather than enlarging the list of the open problems, concludes our Lecture Note by three free speculations on three key issues, selected according to the authors' bias, related to the conceptual framework within which the modeling approach should be developed.

These topics are treated, in the following, as answers to the following three key questions.

KQ1: What should be the relevant key features to account for in the modeling of human crowds?

KQ2: Can some reasonings be developed with the use of artificial intelligence devices on the modeling of crowds?

KQ3: How can the study crowd dynamics contribute to improve the modeling and simulation approach in parallel research fields?

The answer to these questions does not claim to be exhausting, but simply reflects the authors' personal ideas which deserve to be enriched by those of the reader who might be interested to develop them as a research perspective. In doing so, we can observe that all key questions identify some interconnected key problems and, ultimately, indicate a research strategy as stated in a closure which completes our Lecture Note.

• **Key question 1:** The overall contents of this Lecture Note has often asked to account for the heterogeneous individual behaviors of walkers in reaction to interactions with other walkers as well as to the specific features of the areas where the crowds move. Several models do not even consider this aspect and propose a description of walker behaviors as rational entities who homogeneously share the decision process which leads to the walking strategy.

Social behaviors are taken into account by some models where, these behaviors, which can even be not rational, are the same for all walkers. An interesting example introduces even the concept of aggressive walkers that might appear in specific circumstances [26]. However, it is only one possible example out of many.

A key feature of the modeling approach consists of accounting for the heterogeneous behavior of walkers as certain behaviors are not, in practice, equally shared. In addition, certain behaviors do not correspond to a constant state, but are modified by all types of interactions. Some pioneering approaches to account for heterogeneity and social behaviors have been recently taken into account in some models; see for example [12]. However, a systems approach suitable to include a broad variety of possible behaviors is still missing. The assessment of a variety of specific behaviors should be followed by models which depict their dynamics and propagation in time and space.

- **Key question 2:** Devices of artificial intelligence can be used in the modeling of crowds to describe the complex process by which walkers organize their motion. A pioneer paper [28] introduced this topic which focuses on the key problem of the modeling approach. However, we cannot naively claim that a database of observed behavior leads straightforwardly to simulation. Various objectives should be chased to obtain the aforementioned achievement. A minimal choice is as follows.

  – The first target consists in understanding the specific features of individual behaviors during walking dynamics corresponding to different types of interactions. This task needs empirical data devoted to it. Recent literature shows a growing interest in developing this type of investigation, for example [18, 19], and also [29], without forgetting the pioneer paper [21]. A database collecting all validated achievements appears to be an interesting objective within the framework of the theory of databases and their use in artificial intelligence devices [20, 36].

  – The second objective consists in modeling, based on empirical data, a number of behaviors corresponding to the aforementioned empirical investigation. Subsequently, a metric should be defined first to select the empirical data close to the dynamics object of modeling. Finally, the selection of empirical data should be inserted into the model.

- **Key question 3:** The modeling approach to crowd dynamics can referr to the more general framework of the study of self-propelled particles, where the mathematical structures underlying the derivation of models show important common features for a variety of living systems constituted by several interacting living entities, e.g., vehicular traffic, animal swarms, and multicellular systems. These systems differ in the way the said entities, viewed as a-particles, interact among themselves and with the external environment. The differences across systems generate specific models although they referr to the same structure which is deemed to capture the complexity features, at least the most important ones, of multi-agent living systems. Therefore, the research

Figure 5.1: Crowds and swarms.

activity, theoretical and empirical, in a certain field can contribute to understanding how a-particles interact in other systems.

A common feature of driver-vehicles, pedestrians, and animals in swarms is that interactions, and hence collective dynamics, depend on the emotional or social state of the a-particles. However, the criteria for subdividing the overall system into functional subsystems and the expression of heterogeneity differs, e.g., fast and slow cars, trucks, etc. in vehicular traffic and different walking strategies in the case of crowds, leaders, and followers in the case of swarms. Within each functional subsystem heterogeneity appears, for instance drivers' skills in vehicular traffic, walking ability in crowds, and level of stress in animal swarms.

Our reasonings can move also to multicellular systems, where the subdivision into functional subsystems is related by the biological functions expressed by each cell populations, while heterogeneity depends on the level of progression by which cells express their functions. The additional difficulty is that multicellular systems might be not number conservative due to competition and selection.

Finally, we conclude this Lecture Note by focusing on very brief reasonings on future activity on multi-agent systems:

The awareness, by researchers active in the modeling of multi-agent systems, to the need of introducing heterogeneous behavioral features in modeling interactions is rapidly growing and it will promote research activity in the next decade. These indications cannot be confined only to the modeling by kinetic theory methods. It should promote, and the authors of this Lecture Note, believe it will effectively promote, models at each scale within a multiscale vision.

5.6 BIBLIOGRAPHY

[1] B. E. Aguirre, D. Wenger, and G. Vigo, A test of the emergent norm theory of collective behavior, *Social Forum*, 13:301–311, 1998. DOI: 10.1023/A:1022145900928 73

[2] G. Ajmone Marsan, N. Bellomo, and L. Gibelli, Stochastic evolutionary differential games toward a systems theory of behavioral social dynamics, *Mathematical Models and Methods in Applied Sciences*, 26(6):1051–1093, 2016. DOI: 10.1142/s0218202516500251 74

[3] B. Aylaj, N. Bellomo, N. Chouhad, and D. Knopoff, On the Interaction between soft and hard sciences: The role of mathematical sciences, *Vietnam Journal of Mathematics*, 2020. DOI: 10.1007/s10013-019-00381-3 74

[4] B. Aylaj, N. Bellomo, L. Gibelli, and A. Reali, On a unified multiscale vision of behavioral crowds, *Mathematical Models and Methods in Applied Sciences*, 30:1–22, 2020. DOI: 10.1142/S0218202520500013 72

[5] N. Bellomo and A. Bellouquid, On multiscale models of pedestrian crowds from mesoscopic to macroscopic, *Communication a Mathematical Sciences*, 13(7):1649–1664, 2015. DOI: 10.4310/cms.2015.v13.n7.a1 73

[6] N. Bellomo, A. Bellouquid, J. Nieto, and J. Soler, On the multiscale modeling of vehicular traffic: From kinetic to hydrodynamics, *Discrete Continuous Dynamical Systems Series B*, 19:1869–1888, 2014. DOI: 10.3934/dcdsb.2014.19.1869 73

[7] N. Bellomo, D. Clark, L. Gibelli, P. Townsend, and B. J. Vreugdenhil, Human behaviours in evacuation crowd dynamics: From modelling to big data toward crisis management, *Physics of Life Reviews*, 18:1–21, 2016. DOI: 10.1016/j.plrev.2016.05.014 75, 76

[8] N. Bellomo and L. Gibelli, Toward a behavioral-social dynamics of pedestrian crowds, *Mathematical Models and Methods in Applied Sciences*, 25:2417–2437, 2015. DOI: 10.1142/S0218202515400138

[9] N. Bellomo and L. Gibelli, Behavioral crowds: Modeling and Monte Carlo simulations toward validation, *Computers and Fluids*, 141:13–21, 2016. DOI: 10.1016/j.compfluid.2016.04.022 73

[10] N. Bellomo and L. Gibelli, Behavioral human crowds, *Crowd Dynamics Volume 1: Theory Models and Safety Problems*, L. Gibelli and N. Bellomo, Eds., pages 1–14, Birkhäuser, Springer Nature, 2018. DOI: 10.1007/978-3-030-05129-7_1 75

[11] N. Bellomo and L. Gibelli, From "Crowd Dynamics Volume 2" to research perspectives, *Crowd Dynamics Volume 2: Theory Models and Safety Problems*, L. Gibelli and N. Bellomo,

Eds., pages 1–14, Birkhäuser, Springer Nature, 2018. DOI: 10.1007/978-3-030-50450-2 75

[12] N. Bellomo, L. Gibelli, and N. Outada, On the interplay between behavioral dynamics and social interactions in human crowds, *Kinetic and Related Models*, 12:397–409, 2019. DOI: 10.3934/krm.2019017 74, 78

[13] R. Borsche, A. Klar, and F. Schneider, Numerical methods for mean-field and moment models for pedestrian flow, Chapter 7 in *Crowd Dynamics, Volume 1: Theory, Models, and Safety Problems, Modeling and Simulation in Science, Engineering, and Technology*, Birkhäuser, New York, 2018. DOI: 10.1007/978-3-030-05129-7_7 76

[14] D. Burini and N. Chouhad, Hilbert method toward a multiscale analysis from kinetic to macroscopic models for active particles, *Mathematical Models and Methods in Applied Sciences*, 27:1327–1353, 2017. DOI: 10.1142/s0218202517400176 73

[15] D. Burini and N. Chouhad, A Multiscale view of nonlinear diffusion in biology: From, cells to tissues, *Mathematical Models and Methods in Applied Sciences*, 29:791–823, 2019. DOI: 10.1142/s0218202519400062 73

[16] D. Burini, S. De Lillo S., and L. Gibelli, Stochastic differential "nonlinear" games modeling collective learning dynamics, *Physics of Life Review*, 16(1):123–139, 2016. 74

[17] M. Colangeli, A. Muntean, O. Richardson, and T. Thieu, Modelling interactions between active and passive agents moving through heterogeneous environments, Chapter 8 in *Crowd Dynamics, Volume 1: Theory, Models, and Safety Problems, Modeling and Simulation in Science, Engineering, and Technology*, Birkhäuser, New York, 2018. DOI: 10.1007/978-3-030-05129-7_8 75

[18] A. Corbetta, L. Bruno, A. Mountean, and F. Yoschi, High statistics measurements of pedestrian dynamics, models via probabilistic method, *Transportation Research Proceedings*, 2:96–104, 2014. DOI: 10.1016/j.trpro.2014.09.013 78

[19] A. Corbetta, A. Mountean, and K. Vafayi, Parameter estimation of social forces in pedestrian dynamics models via probabilistic method, *Mathematical Biosciences Engineering*, 12:337–356, 2015. DOI: 10.3934/mbe.2015.12.337 78

[20] H. De Sterck and C. Johnson, Data science: What is it and how is it thought?, *SIAM News*, 48:1–6, 2015. 78

[21] D. Helbing and P. Molnár, Social force model for pedestrian dynamics, *Physical Review E*, 51:4282–4286, 1995. DOI: 10.1103/physreve.51.4282 78

[22] D. Hilbert, Mathematical problems, *Bulletin American Mathematical Society*, 8(10):437–479, 1902. 73

[23]   A. R. Karlin, *Game Theory, Alive*, American Mathematical Society Press, 2017. DOI: 10.1090/mbk/101 74

[24]   M. Kinateder, T. D. Wirth, and W. H. Warren,   Crowd dynamics in virtual reality, Chapter 2 in *Crowd Dynamics, Volume 1: Theory, Models, and Safety Problems*, *Modeling and Simulation in Science, Engineering, and Technology*, Birkhäuser, New York, 2018. DOI: 10.1007/978-3-030-05129-7_2 76

[25]   H. R. Kwon and E. A. Silva, Mapping the landscape of behavioral theories: Systematic literature review, *Journal of Planning Literature*, Article Number: UNSP 0885412219881135, 2019. DOI: 10.1177/0885412219881135 73

[26]   Y. Li, M. Chen, X. Zheng, Z. Dou, and Y. Cheng, Relationship between behavior aggressiveness and pedestrian dynamics using behavior-based cellular automata model, *Applied Mathematics and Computation*, 371:124941, 2019. DOI: 10.1016/j.amc.2019.124941 78

[27]   J. Lin and T. A. Luckas,   A particle swarm optimization model of emergency airplane evacuation with emotion, *Networks Hetherogeneous Media*, 10:631–646, 2015. DOI: 10.3934/nhm.2015.10.631 76

[28]   Y. Ma, E. Wai Ming Lee, and R. Kwok Kit Yuen, An artificial intelligence-based approach for simulating pedestrian movement, *IEEE Transactions on Intelligent Transportation Systems*, 16(11):3159, 2016. DOI: 10.1109/tits.2016.2542843 78

[29]   M. Moussaid, D. Helbing, S. Garnier, A. Johanson, M. Combe, and G. Theraulaz, Experimental study of the behavioral underlying mechanism underlying self-organization in human crowd, *Proc. Royal Society B: Biological Sciences*, 276:2755–2762, 2009. 78

[30]   M. A. Nowak, *Evolutionary Dynamics. Exploring the Equations of Life*, Harvard University Press, 2006. DOI: 10.2307/j.ctvjghw98 74

[31]   E. Ronchi,   Disaster management: Design buildings for rapid evacuation,   *Nature*, 528(7582):333, 2015. DOI: 10.1038/528333b 76

[32]   E. Ronchi, F. Nieto Uriz, X. Criel, and P. Reilly,   Modelling large-scale evacuation of music festival, *Fire Safety*, 5:11–19, 2016. DOI: 10.1016/j.csfs.2015.12.002 76

[33]   E. Ronchi and D. Nilsson, Pedestrian movement in smoke: Theory, data, and modelling approaches, Chapter 3 in *Crowd Dynamics, Volume 1: Theory, Models, and Safety Problems*, *Modeling and Simulation in Science, Engineering, and Technology*, Birkhäuser, New York, 2018. DOI: 10.1007/978-3-030-05129-7_3 75

[34]   A. Schadschneider, M. Chraibi, A. Seyfried, A. Tordeux, and J. Zhang, Pedestrian dynamics: From empirical results to modeling, in *Crowd Dynamics Volume 1: Theory Models*

and Safety Problems, L. Gibelli and N. Bellomo, Eds., pages 63–102, Birkhäuser, Springer Nature, 2018. DOI: 10.1007/978-3-030-05129-7_4 76

[35] L. Wang, M. Short, and A. L. Bertozzi, Efficient numerical methods for multiscale crowd dynamics with emotional contagion, *Mathematical Models and Methods in Applied Sciences*, 27:205–230, 2017. DOI: 10.1142/s0218202517400073 74

[36] "Web Source," OECD, Organization for Economic Co-Operation and Development, Paris, France. *Data-Driven Innovation, Big Data for Growth and Well-Being*, OECD Publishing, www.oecd.org/sti/ieconomy/data-driven-innovation, 2015. 78

[37] N. Wijermans, C. Conrado, M. van Steen, C. Martella, and J. L. Li, A landscape of crowd management support: An integrative approach, *Safety Science*, 86:142–164, 2016. DOI: 10.1016/j.ssci.2016.02.027 76

Authors' Biographies

BOUCHRA AYLAJ

Bouchra Aylaj is an Associate Professor with Habilitation in mathematics at University of Hassan II of Casablanca, Faculty Ain Chock, Morocco. She started her career in 2006 when she was called to develop a research program on mathematical modelling in biology. Her scientific activity has been focused on the following topics: scientific computing and control for risk analysis and analytical and computational problems in epidemiology. Subsequently, she moved her scientific interests to the modeling and related safety problems focused on social behaviors in human crowds.

NICOLA BELLOMO

Nicola Bellomo is a distinguished professor at the University of Granada and Professor Emeritus at the Polytechnic University of Torino. He started his career in 1980 when he was called to cover the chair of mathematical physics and applied mathematics due to his scientific achievements on the mathematical theory of the Boltzmann equation and of stochastic differential equations. Subsequently, he moved his scientific interests to the study of living systems, becoming one of the pioneers of the development of active particles methods to the modeling of large systems of self-propelled interacting entities. He is author of two books published by Birkhäuser devoted to this topic. In 2009, he delivered the prestigious Shank Lecture on the modeling of immune competition, and was awarded the "Third Level Honor" in 2016 for scientific merits by the President of the Italian Republic.

LIVIO GIBELLI

Livio Gibelli is a Lecturer in Mechanical Engineering at the University of Edinburgh. He received his Ph.D. in applied mathematics from the Politecnico di Milano and, prior to the current position, he worked as Research Fellow at the University of Warwick, Politecnico di Milano, Politecnico di Torino, and University of British Columbia. His main research interests include non-equilibrium multiphase fluid flows, the continuum description of slightly rarefied gases, the numerical methods for solving kinetic equations, and the modeling of crowd dynamics.

DAMIÁN KNOPOFF

Damián Knopoff is a chemical engineer and mathematician, holding a Ph.D. in Mathematics from Cordoba National University. Currently, he is an Associate Researcher at the Argentinian Scientific and Technical Research Council. His main research fields include nonlinear dynamical systems and numerical methods for differential equations with applications to the modeling and simulation of complex living systems, including biological phenomena, socio-economic systems, and crowd dynamics.